CONSTRUCTIVE
CONTROL

CONSTRUCTIVE CONTROL

CONTROL

Design and Use of
Control Systems

William H. Newman

Samuel Bronfman Professor
  of Democratic Business Enterprise
Graduate School of Business
Columbia University

PRENTICE-HALL, INC. / *Englewood Cliffs, N.J.*

Library of Congress Cataloging in Publication Data

Newman, William Herman, 1909-
    Constructive control.

    Includes bibliographies.
    1. Industrial management.   I. Title.
HD38.N39   658.4   74-13306
ISBN 0-13-169367-0
ISBN 0-13-169359-X (pbk.)

Printed in the United States of America

10   9   8   7   6   5   4   3   2

PRENTICE-HALL INTERNATIONAL, INC. London
PRENTICE-HALL OF AUSTRALIA, PTY. LTD.   Sydney
PRENTICE-HALL OF CANADA, LTD.   Toronto
PRENTICE-HALL OF INDIA PRIVATE LIMITED   New Delhi
PRENTICE-HALL OF JAPAN, INC.   Tokyo

to

Edgar M. Bronfman
*   perceiver of new ways*
*   for excellence in management*

CONTENTS

II

ADAPTING DESIGN
TO DIFFERENT SITUATIONS

10

INTEGRATING CONTROLS WITH TOTAL MANAGEMENT DESIGN 144

11

PUTTING AN IMPROVED SYSTEM TO WORK 162

PREFACE

THE MORE DYNAMIC—or topsy-turvy—our world becomes, the greater the need for well designed controls. Yet, in fact, improvements in managerial control have been neglected. We have been so busy updating plans and modifying organizations that the closing phase of the managerial cycle—control—has received comparatively little managerial attention. Consequently, opportunities abound for reshaping controls to fit current needs.

The past neglect of control arises partly from a widespread feeling that controls are always punitive and personally degrading. Actually, when we think of the crucial role of control in landing a man on the moon or in directing a great orchestra, the concept of control shifts to a normal, constructive process. Controls can and should be designed to prompt this latter kind of positive response.

This book outlines a fresh approach to managerial control that stresses positive, future-oriented, behavioral response. Suggestions are made for broadened measurement and feedback to achieve this end. Three facets are stressed:

(a) The impact of controls on *future* action.
(b) The *behavioral response* of people, necessary to make controls effective.
(c) Tying controls to *desired results*, both tangible and intangible.

This constructive, forward-looking view of managerial control is developed in three stages. Part I lays out the structural and behavioral *elements* that should be considered in every control cycle. Part II explores the *application* of these concepts to a variety of situations, starting with repetitive operations, examining projects and resource control, and extending to creative activities and strategy. The balancing of various controls in a total structure and the integrating of controls with other phases of management are considered in Part III on control *systems*.

The framework presented will be useful to managers who design and exercise controls—and to pre-managers who are developing
ideas about how to sail better when they are at the helm. Although
I have reviewed scholarly discourses and behavioral science reports
dealing with control, the inevitable jargon and abstraction of such
writings have been translated into language readily grasped by an
experienced manager. On the other hand, descriptions of detailed
control techniques have also been avoided. To keep the book fairly
short and to focus on its major thrust, many excellent techniques for
specific applications are only referred to in the suggested readings.
My hope is that both the scholar and the manager will be able to
move quickly from concepts presented here to a constructive redi-
rection of his own efforts.

This study was made possible by a continuing grant of the
Samuel Bronfman Foundation to the Graduate School of Business
at Columbia University. I also had the good fortune of insightful
research assistance by Raymond-Alain Thietart. And the final conversion of ideas to manuscript reflects the patience and care of
Camilla Koch. For this help I am most grateful.

W. H. N.

CONSTRUCTIVE CONTROL

I
BASIC CONCEPTS

chapter I

ROLE OF CONTROL
IN PURPOSEFUL ENDEAVORS

LONG pushed aside by the excitement of planning and the challenge of organizing, refinements of controlling have received comparatively little attention by managers. Today, managerial control offers great opportunities for improving the effectiveness of private and public enterprises.

A fresh, uninhibited attitude is necessary. Too many people think of managerial control as only a repressive, negative activity. Like taxes, control is often regarded as an unavoidable burden. At the same time, we marvel at controls which enable our astronauts to land on the moon, right on target and on time. And we find the scorekeeper a convenient fellow at the ball game. The negative viewpoint arises in large part because of inadequate design of managerial controls and inept use of those we have.

Constructive, Future-oriented Control

To grasp the full potential of managerial control, we need a broad, constructive view of the control process. Our attitudes and expectations should embrace the following points.

1. *Control is a normal, pervasive, and positive force.* Evaluation of results accomplished and feedback of this information to those who can influence future results is a natural phenomenon. The cook watches the pie in the oven; the orchestra conductor listens to his orchestra—and its recordings; the doctor checks his patient; the oil refiner tests the quality of his end-product; the farmer counts his chickens; the football coach keeps an eye on the scoreboard.

The news received may be good or bad, and the "corrective action" may be encouragement or restraint. Assuming a purpose or goal, each person and manager needs to know what progress he is making. There is nothing sinister nor dictatorial about such controlling. Rather, it is a normal aid in achieving results.

2. *Managerial control is effective only when it guides someone's behavior.* Behavior, not measurements and reports, is the essence of control. We often become so involved with the mechanics of control that we lose sight of its purpose. Unless one or more persons act differently than they otherwise would, the control reports have no impact. Consequently, when we think about designing and implementing control, we must always ask ourselves, "Who is going to behave differently, and what will be the nature of his response?"

Some controls provoke over-reaction. Many profit-centered controls, for instance, lead to excessive preoccupation with very short-run results. On the other hand, controls seeking personnel changes —such as increased employment of blacks—often get token acceptance and may even lead to practices that restrict "equal opportunity." It is the behavioral response to controls that really matters.

3. *Successful control is future-oriented and dynamic.* Long before the Apollo spacecraft reached the moon, control adjustments had been made. Similarly, we don't wait until next year's sales are recorded to make adjustments in packaging or pricing which are necessary to achieve the goal; instead, we use early measurements to predict where our present course is leading, and modify inputs to keep us on target. PERT controls (see Chapter 5) on the construction of a new plant are designed to catch promptly delays in early stages which will put off the completion date; then special attention is focused on overcoming the critical slippage.

In each of these examples evaluation comes early and involves prediction. Fine tuning then can be introduced prior to the main completion date. Even those evaluations made after work is completed yield their chief benefit in guiding similar activity in the future.

The future is uncertain. So most controls provide for adjusting to unexpected conditions. For some routine operations we can safely use a static response—a household thermostat is a control of this kind. But control of a research program presumes that new findings will probably call for actions which cannot be laid out in advance. And in designing pollution controls we must anticipate shifts in social standards that we are expected to meet. In such situations the control design should include monitoring of the environment to flag changes in operating conditions. Increasingly, controls must aid managers in reaching a "moving target."

4. *Control relates to all sorts of human endeavors.* The need for evaluation and feedback is just as pressing in charitable organizations as in profit-seeking corporations. Each is concerned with attaining its goals and each has limited resources. Moreover, control should not be confined to easy-to-measure results. The quality of service in a hospital or bank, the training and promotion of minority workers, and the resourcefulness of a purchasing agent in developing alternative sources for important supplies—all need to be controlled.

Ingenuity may be required to devise measures of intangible output. And in non-profit enterprises the tying of control standards to objectives is often complicated by multiple goals. Nevertheless, constructive managerial control has a vital role to play whenever people join their efforts to achieve some common purpose.

Traditional controls miss many opportunities. To obtain the potential benefits we need a fresh approach—viewing control as a positive force, concentrating on behavioral responses, taking a future orientation, and including intangible and long-run results in a balanced control system.

Focus on Evaluation and Feedback

Control is one of the basic phases of managing, along with planning, organizing, and leading. In this book we shall assume that these other phases of managing are performed satisfactorily. With plans laid, necessary resources assembled and organized, and action initiated by competent leaders, control closes the managerial cycle—comparing actual results with plans and contributing to the next round of planning, organizing, etc.

In practice, the phases of managing are closely intertwined. For example, the output of planning provides the basis for control standards; organizing helps determine who should take corrective action; leading sets the tone for participation in selecting pars, or levels of achievement, and for self-control. We shall frequently draw attention to these interdependencies. But our mission here is to help managers attain balanced, constructive control.

The word "control" has diverse meanings, ranging from financial ownership to a throttle on a motorboat, and from hidden power to psychological influence. However, we shall center our attention on managerial control—*the series of steps a manager takes to assure that actual performance conforms as nearly as practical to plan.* Confusion can be avoided by *not* using "control" loosely to embrace power, authority, influence, and leadership. These and other factors contribute to desired results, but if we include all of them in the concept of control the term becomes so broad it loses any clear meaning.

Managerial control always includes evaluation and feedback. So a simple way to check one's thinking is to ask whether these elements are present. The evaluation may occur at various stages and the feedback may go to different people, as we shall see in later chapters. But the cybernetic idea of a process involving (1) appraisal, and (2) opportunity to respond to that appraisal can be found in all control situations.[1]

Basic Types of Control

Control efforts can be made much more effective by recognizing three different types of control:

1. *Steering-controls.* Results are predicted and corrective action is taken before the total operation is completed. For example, flight control of the spacecraft aimed for the moon began with trajectory measurements immediately after take-off and corrections were made days before actual arrival.

2. *Yes-no controls.* Here, work may not proceed to the next step until it passes a screening test. Approval to continue is required. Quality checks and legal approval of contracts are examples.

3. *Post-action controls.* In this type of control action is completed; then results are measured and compared to a standard. The

---

[1] The above definition of control is consistent with the common usage among U.S. managers. There is sufficient variation, however, to warrant careful listening when the term is used. For example, R. N. Anthony finds the distinction between planning and controlling hard to draw, and then compounds confusion by using "management control" to embrace both planning and control which occurs between strategic planning and routine operations. (See *Planning and Control Systems.* Boston: Harvard Graduate School of Business Administration, 1965.) In contrast, A. S. Tannenbaum considers all sorts of influence in his *Control in Organizations* (New York: McGraw-Hill Book Company, 1968). H. Smiddy and his associates at General Electric Company tried to side-step the confusion—and the negative connotation of controlling—by emphasizing "measuring" as one of the four basic management processes (along with planning, organizing, and integrating), but this formulation has not gained general acceptance. The present analysis seeks to enrich and sharpen the meaning of a recognized word rather than use old words in unfamiliar ways or invent a new jargon.

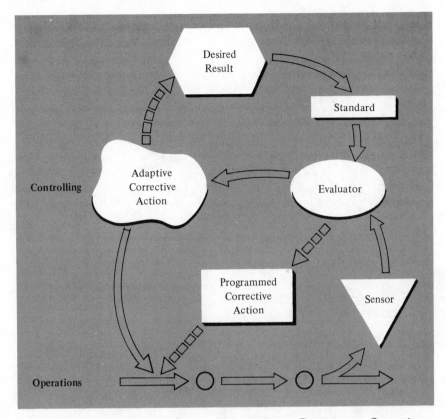

FIGURE 1-1. *Control Cycle for Repetitive or Continuous Operations.*

typical budgetary control and school report cards illustrate this approach.

All three types may be needed to control a department or major activity. But it is steering-controls that offer the greatest opportunity for constructive effect. The chief purpose of all controls is to bring actual results as close as possible to desired results, and steering-controls provide a mechanism for remedial action while the actual results are still being shaped. Much of the discussion in the following chapters will focus on the design of good steering-controls.

Yes-no controls are essentially safety devices. The consequences of a faulty parachute or spoiled food are so serious that we take extra precautions to make sure that the quality is up to specifica-

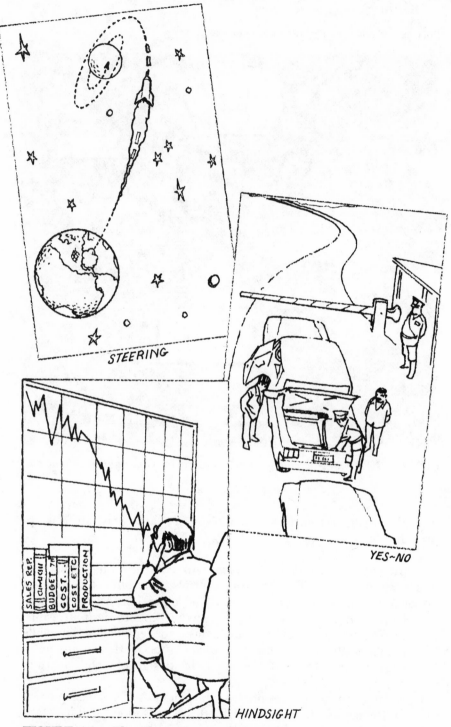

STEERING

YES~NO

HINDSIGHT

FIGURE 1-2. *The Three Types of Control.*

tions. Avoidable expense or poor allocation of resources can also be checked by yes-no controls. If we could be confident that our steering-controls were effective, the yes-no controls would be unnecessary; unfortunately, steering-controls may not be fully reliable, or may be too expensive, so yes-no controls are applied.

Post-action controls, by definition, seem to be applied too late to be very effective. The work is already completed before it is measured. Actually, post-action controls do serve two purposes. (a) If rewards (a medal, bonus, discharge, self-esteem, etc.) based on actual results have been promised, these results must be measured and the appropriate rewards made to build future expectations about the close relationship between actual results and rewards. The aim is psychological reinforcement of the incentive scheme. The pay-off in this reinforcement lies in future behavior. (b) Post-action controls also provide planning data if similar work is undertaken in the future.

The appropriate spots to employ these different types of control will be illustrated throughout the book. Often they can be combined into small control systems. Although the idea of a single, fully integrated system for an entire company is appealing, we shall not propose a quick all-embracing formula. The wide diversity of controls coupled with the benefits of keeping them closely in tune with local needs makes such a single system impractical. Instead, better results are achieved by designing a series of localized systems which are tied together with selected checks on overall results.

Issues Examined in This Book

Whether he likes it or not, a person in a managerial post must be deeply involved in control. Our aim is to help him improve the effectiveness of control efforts related to activities under his direction. The contention is that he can achieve results best not by increasing the time he personally spends controlling, but (1) by carefully designing the controls to be used by persons close to the scene of action, and (2) watching how these evaluation devices actually affect behavior in his organization.

The analysis is divided into three parts:

1. *Basic concepts.* The issues and problems arising time after time in almost any control sequence are examined in Chapters 2 and 3. First, the elements in the design of a control cycle are set

forth, and then human responses to controls are explored. These chapters on the formal and human aspects alert us to the basic components of control design.

2. *Adapting design to different situations.* A major obstacle to good control is the widely different kinds of results and operating situations involved. So a second approach deals with an array of situations from routine and stable to broad and uncertain. Chapters 4 through 8 focus on repetitive operations, projects and programs, resources (including financial budgeting), creative activities, and strategy. In some of these areas control techniques are highly developed, whereas in others we are still pioneering.

3. *Control systems.* No managerial control works in isolation. It competes for attention with other controls and is intertwined with the entire managerial process. This interaction of the parts with the whole is analyzed in Chapters 9 and 10—balancing the total control structure, and integrating controls with the total management design. A concluding chapter summarizes the recommended approach; and discusses briefly the salient problems in making a revised control system work.

Throughout the discussion we shall be dealing with steering, yes-no, and post-action controls—suggesting ways these can be adapted to particular needs and also fitted into a balanced system. Although past results help in such controls—because "what is past is prologue"[2]—our emphasis is on influencing behavior to achieve future results.

FOR FURTHER READING

BEER, S., *Decision and Control.* New York: John Wiley & Sons, 1966.
  For reader who wants to relate cybernetics to managerial control.

JEROME, W. T., *Executive Control.* New York: John Wiley & Sons, 1961.
  Broad view of control as integral part of management process.

MOCKLER, R. J., *The Management Control Process.* New York: Appleton-Century-Crofts, 1972.
  Comprehensive text covering concepts and techniques related to

---

[2] Shakespeare, *The Tempest,* inscription on National Archives Building, Washington, D.C.

managerial control. The frequent references provide leads into control literature.

Murray, W., *Management Controls in Action*. Dublin: Irish National Productivity Committee, 1970.
Clear report on field study of control, illustrating efficiency control, operating control, and strategic control.

chapter 2

ELEMENTS IN DESIGN
OF A CONTROL CYCLE

THE design of every control cycle involves a series of elements. Each of these elements is always present in steering-controls, and most are found in yes-no and post-action controls. An awareness of these elements helps us avoid naive arrangements which may create as much mischief as they do good. So, when designing a control cycle, the following questions should be answered.

1. What are the dimensions of *desired results,* and how should such aims be expressed so that control is aided?
2. Can *predictors* of these results be established at early stages?
3. What combination of predictors will provide a sensitive, reliable, balanced, and economical *composite* feedback?
4. What *par* is desired for each characteristic measured?
5. How should the resulting *information flow*—that is, what is reported to whom, when, and how?
6. For non-routine results, who *evaluates*—who decides on significance, identifies causes, and projects consequences of unmodified behavior?
7. How is *corrective action* decided upon and put into motion?

Regardless of the subject of control—our household drinking water, broadcast of a TV show, production of a steel ingot, air pollution, or winning an election—these questions are pertinent. In this chapter, the nature of each element and several options available to the designer will be examined. The behavioral responses to various designs will be explored in Chapter 3. With these basic concepts in mind, we can then turn in Part II to their application to various kinds of managerial objectives.

Define Desired Results

Control always starts with an objective. We want a particular set of events or conditions to happen in the future.

Express results in measurable dimensions, if possible. The objective may be as clear-cut as "All mail delivered by December 25th." But often our objective is vague—"Good delivery service," for example, or "Fair treatment of dispossessed families." The trouble with such vague objectives for control purposes is that we really don't know what to measure and to push for, and we are not sure when we have achieved success.

Most objectives can be defined in measurable terms, if we put our minds to it. The dimensions of "Good delivery service," to continue that example, include arrival at the customer's door when promised, no damage to the product, placement convenient for the customer, courtesy, and perhaps demonstration or instruction. The dimensions of "Fair treatment for dispossessed families" are more controversial; but if we are to develop meaningful control we need

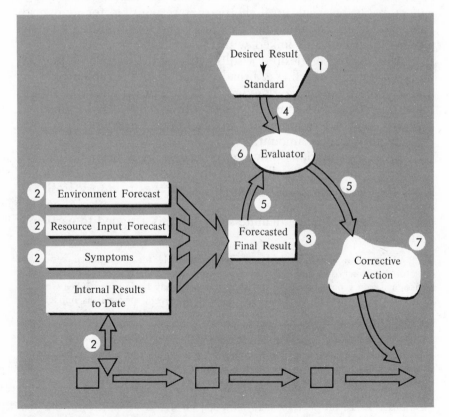

FIGURE 2-1. *Steering-Control.*

to identify such features as length of advance notice, finding at least one available comparable dwelling, and financial aid for moving expenses.

Such dimensions need not be static. As we gain experience, the criteria can be refined, and perhaps new facets added. When objectives change, a corresponding adjustment in stipulated results should, of course, be made. But to design even an informal control mechanism we do need a clear understanding of just what results are sought.

Link results to individuals. In addition to expressing results in sharp, measurable dimensions, results should be tied to individuals. Managerial control, as we shall stress repeatedly, seeks to influence the behavior of people. Consequently, we need to link particular desired results to those individuals who can most affect the outcome. Thus, control over the treatment of dispossessed families will probably be weak until the persons responsible for giving advance notice, for finding other dwellings, and for allocating financial aid are identified and linked with the control objective.[1]

This mating of desired results with particular people may require either (1) dividing a broad objective into sub-objectives, or (2) adjusting the organization, or (3) both. For instance, the underlying concept of "responsibility accounting" is to classify incomes, expenses, and assets according to the organization unit that most directly affects their amount.[2]

In many situations, however, clarification is needed. The objective of control over pollution in the Hudson River is too broad. Similarly, in an ice cream plant, to "keep the inventory fresh but adequate" is too vague a goal. In both cases control becomes operational only when a definite set of goals is assigned to particular individuals.

[1] More than one individual may be accountable for a particular result: an operator, his supervisor, and the supervisor's supervisor, for instance; or a task may be assigned to a group; both line and staff personnel may be involved. (See Chapters 9 and 10.) Even though a control may impinge on several persons, its impact need not be ambiguous. There can still be a clear recognition of which individuals are in positions to exert the greatest influence on the outcome.

[2] This linking of desired results with individuals is closely related to company planning and organization. *If* the planning process within our company has already generated long- and short-range goals, and amplified these into sub-goals and sub-sub-goals, then the objectives sought by control are already set. Also, clearly defined organization helps by indicating who is to contribute what in the achievement of these goals. Especially in companies operating on a "management by results" approach, the end objective of control activity and the dimensions may be readily ascertained. Under such circumstances we can proceed in our control design to questions 2 through 7 listed above.

A word of caution. Defining desired results (1) in clear dimensions that are measurable and (2) in clusters that can be linked to particular individuals will indeed aid the process of controlling. But we must not limit our attention to what is easy. Some results of a company's activities are hard to pin down, yet they may be fully as important as results which are easy to measure. Such features as creativity or community relations (which are discussed in later chapters) are often essential parts of balanced performance, and we must include them in our total management design even though the best control mechanisms we can devise are messy. It is desired results that dictate what is to be controlled. We don't drop the drummer and cellist from the orchestra because their instruments are hard to carry.

Look for Predictors of Results

Value of predicted results. The idea that control consists only of "comparing actual results with desired results" is both wrong and dangerous. By the time we know actual results, it's usually too late to do anything about them. We don't wait until an Apollo moonshot does or does not hit its target. Long before the actual event occurs, we *predict* the result of continuing on our present course, and corrective action is based on those predictions.

Effective managerial control is largely based on predictions of results, not actual results. If a cholera epidemic, an electric power failure, an uninspired symphony performance, an inflationary price rise, a wage bill 20 percent over budget, or some other off-target result actually occurs, the time for control action has already passed. What's done is done. We cannot reverse history. Instead, we try to anticipate what is likely to happen, and on the basis of these forecasts—these predictions—we take corrective action which we hope (predict) will help bring actual results close to the control target.

Control is future-oriented; otherwise, it is futile.

Early warning predictors. Since effective control requires prediction of actual results, a crucial element in each control design is finding one or more good predictors. In practice, such predictors, or sensing devices, are as diverse as the desired results they forecast. The following list merely suggests the array of possibilities.

1. *Input Measurements.* The level of some key input may predict the actual results of an operation. For example, orders received foretell shipments at a later date; inquiries from customers often

predict orders; a knowledge of the level of a customer's inventories may forecast inquiries. Payroll changes provide insights into future labor expense. Purchase contracts for parts afford cues about future costs and shipping dates.

2. *Success of Early Steps.* The ease or difficulty of completing early steps and the revelation of unanticipated difficulties are frequently used to predict final results. Thus, the completion of engineering specifications and other implementation plans provide a basis for revising the estimated cost and delivery date of a nuclear power plant. Early reviews of a play or a book, like the initial reception of a new product, give strong clues to longer-run popularity.

3. *Sophistication of the Process.* The degree of sophistication and skill with which an operation is conducted provides another basis for predicting results. In schools, for example, preparation of clear lesson plans throws some light on the likelihood of a good learning experience. The number of "cold calls" by an insurance agent has a bearing on the volume of business he is likely to write during the following months. In a food plant the attention to moisture content foretells the quality of cornflakes that will be produced the next day. In observing how any process is performed, we try to identify those variables which are most likely to "cause" differences in final results.

4. *Symptoms.* Although symptoms do not directly affect actual results, they can be useful predicting devices. Tardiness and absenteeism reflect worker morale and may forecast low labor efficiency. Even more subtle, the vibrations of his building enable one production vice president to sense machine shutdown and production delays long before he receives reports of trouble. Traffic noise helps an urban hostess predict whether her guests will be late or on time.

5. *Assumed Conditions.* All expectations of desired results are based on assumed operating conditions, such as high government spending, no strikes, expansion of competitors' capacity, and the like. Careful monitoring of these key assumptions may indicate an external change which will help or hinder the achievement of results. Here, attention is on necessary conditions.

Considerable ingenuity is needed in finding good predictors for use in control. Obviously, the sooner—and the more reliably—a gap between desired results and predicted results is recognized, the greater will be the opportunity to take corrective steps.

Observation of actual results.   This future orientation of managerial control does not mean that we give no attention to the actual results that do occur. A comparison of final results with desired results (i.e., post-action control) may be useful for one or more of the following reasons.

1. In a continuing flow of repetitive operations—such as producing shoes or processing social security payments—evaluation of today's results may suggest changes for tomorrow's operations.

2. When somewhat comparable (but not identical) activities will be undertaken in the future, comparison of actual results against targets may suggest an opportunity to improve the process that will be used in the next cycle. For example, a failure of last year's sales to come up to budget may suggest more advertising next year. But note that in both the above situations we are using an assessment of the actual results from past efforts to signal the need for somewhat different efforts in the future. The payoff comes from improved operations in the future.

3. If rewards or penalties based on actual performance have been promised, then measurement of final results should be made and the appropriate rewards given. This builds future expectations about the close tie between actual results and rewards. Psychological reinforcement of the incentive scheme occurs. And here, again, the payoff from the reinforcement lies in future behavior.

The proposition that managerial control is primarily concerned with predictions and then modifications of future results needs one qualification. Yes-no controls are screening devices. Actual results to date are measured and work is permitted to proceed only if the results are satisfactory. Test flights of aircraft, quality checks of food to be processed, and reviews of a legal contract are examples. If results to date are found to be unsatisfactory, a correction must be made. (Or if several units are presented, only the acceptable ones are forwarded to the next step.) Such screening-controls, although vital in some situations, are much less common than predictive controls. In later chapters we shall illustrate when yes-no controls are needed.

Select Composite Feedback

With luck and perseverance we will uncover several different ways of predicting what is likely to occur in each of the desired result areas. The question then arises which of the available predictors to use? Since one predictor may give us early but erratic signals, whereas another warns of external trouble, and a third is late but comprehensive, a composite of several feedbacks is often needed for balanced control of a key target.

Guides for selecting predictors. Several characteristics are desirable in a composite feedback scheme.

1. *Promptness* is critical. E. C. Schleh concludes, "Above all a management control system must be current." [3] The possibility of successful corrective actions with a minimum of resistance is enhanced when the need for adjustment is discovered before a lot of personal effort has been invested and before current facts are forgotten.

2. *Reliability*, although subordinate to promptness, is obviously desirable. False leads waste executive effort. So, when possible, a predictor which occasionally is erratic should be coupled with some kind of confirmation test. Thus, when the small-craft warning goes up, the experienced sailor keeps a closer watch on the movement of clouds.

3. *Coverage* also warrants attention. Perhaps a long queue of customers, for example, flags serious inventory and production problems; or a low number of grievances indicates satisfied workers. But such predictors give only a partial picture; other aspects of the total situation must be checked. A more comprehensive review may occur less frequently or with a lag, but it does help maintain balanced effort. With a narrow index, there is always the temptation to manipulate just that one factor.

4. *Expense* of measuring varies widely. An index of new orders may be easily compiled from records that are kept for other purposes. Field interviews with users of a product are much more expensive. And the costs of testing pharmaceuticals for side effects in human beings rises to still another order of magnitude.

All these considerations—promptness, reliability, comprehensiveness, and expense—lead one sales manager to use a composite of customer inquiries, shipments, field visits, and order backlog to assess sales results.

Use of sampling. Coupled with the task of picking a good combination of predictors is deciding where sampling is appropriate.

[3] *Management by Results.* New York: McGraw-Hill Book Company, 1961, p. 179. Similarly, G. S. Odiorne says, "Prompt feedback is far more important in changing behavior than intensity of feedback." (*Management by Objectives.* New York: The Macmillan Company, 1973, p. 167.)

The expense of measuring can be greatly reduced, for instance, if we compute the speed of serving hospital patients for only one out of one hundred such patients. Or, the president of Marshall Field's department store may read only a random selection of twenty-five of all complaint letters received each week. The occasional visit of an executive to a branch office is another form of sampling. Although such sampling admittedly reduces the reliability of the measurement, any major opportunity for improvement hopefully will be detected.

In a highly repetitive operation in which hundreds of similar actions occur, sampling can be done on a very sophisticated basis and the significance of deviations can be computed statistically. Such statistical sampling was first applied to quality control of mass-produced items, and the technique has been extended to credit activities in mail-order companies and to the processing of income tax returns. The statistical analysis coupled with sampling reduces both the expense of measuring and the expense of investigating deviations.

Strategic control spots. Too much control data can smother the whole system. More than the expense of compiling the data is involved. Busy executives have to allocate valuable time to reviewing numerous reports; they may become preoccupied with minor discrepancies; and then in reaction to the heavy burden, some will disregard the entire mass of data thrust upon them. The art is to pick strategic control spots—just a few predictors and screens that provide adequate warnings and checks, but not too many.

One executive receives a sample of incoming mail; another watches a chart of advanced bookings; a nineteenth-century meat-packer is reputed to have kept an eye on the stream that carried off the waste from his plant. Such petty control points are likely to be too narrow, as our previous discussion of desired results and predictors implies. Nevertheless, parsimony in the number of feedbacks has its virtue. The best control systems are often simple. A few indicators are watched closely, with other data readily available when further investigation is signaled.

Set Par for Each Predictor and Desired Result

Identifying the factors and indexes to watch for purposes of control still leaves us with the question of "How good is good?" What level of achievement is acceptable?

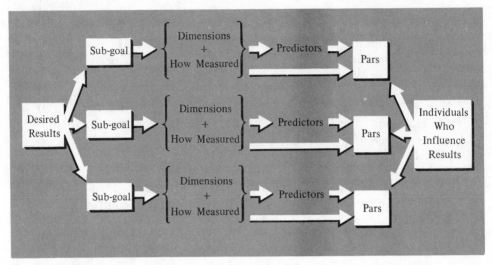

FIGURE 2-2. *Designing Control Standards.*

When is the signal up? To know when corrective action is, or is not, called for we need a par for each predictor as well as an acceptable par for each characteristic of the desired results. Thus, in public health an occasional case of hepatitis may be normal, but three or more cases in one week is a warning of a potential epidemic. For a person, more than one degree deviation in temperature above or below the 98.6 degrees par is a symptom of trouble.

On highways, the speed limit is a single standard that is applicable to all drivers. Similarly in business, a single standard may be applicable to such things as the quality of delivery service. But for many other characteristics we may adjust the standard for a particular individual and local circumstances. Thus we may say that 3,000 encyclopedias is par for export shipments in the second quarter (down 7 percent from the first quarter), or sixty voter registrations is par for Sally Hardin (ten above the average). Adjusting pars enables us to fit a control mechanism to a particular circumstance.

Securing flexibility through adjustment of pars. In controls, as in other phases of management, we have legitimate need for flexibility. For instance, a firm may increase inventories if there is reason to anticipate a shortage of raw materials; it may cut its price to meet competition knowing that the dollar sales figure will be thrown out of line; during a depression it may decide not to cut employment in proportion to the drop in production; and so forth.

Although such changes do make previous pars inappropriate, control is still feasible. The sensible way of dealing with unforeseen conditions is to adjust par. The performance characteristics being watched, and their predictors, continue to be useful; we need to change only the *levels* of expectation.

Specify Information Flow

Control data must be communicated to the people who can act on it. Typically, measurements and comparisons with pars are made by clerks or specialists or measuring devices. Immediately a question arises—what to report, to whom, when, and how? Spectacular news will spread rapidly over informal networks, but most control data will flow through the channels we establish.

Communication network. Where to send information depends on the local organization design. Nevertheless, a few general suggestions apply to most situations, or at least raise issues to consider.

Always feed back data promptly to the person making operating decisions. He is the one most likely to be able to respond with positive action, and the sooner he acts the better. Besides, we want to promote his sense of responsibility for achieving the best results possible; if the control report indicates that he needs help, encourage him to ask for it—but never bypass him.

The supervisor of the local decision-maker should also receive control data. But frequently summarized information at a later time is adequate since the supervisor's primary role is to make sure that the local person has already responded to the report. However, signs of a major opportunity or crisis should go both to the local person and his supervisor concurrently; the supervisor will probably help decide how to respond to such situations and so get him into the act promptly.

The flow of control data to staff depends on what role we expect the staff to play. Usually staff is more concerned with planning and detailed studies of specific problems; consequently, they do not need control reports on a current basis. Another arrangement, however, is that the staff does the measuring and makes a technical evaluation, or the staff may be very active in coordination, in which cases they too need data promptly. A sure way to create friction is to send control data to staff first so that they can call for remedial action before the operating people have an opportunity to respond on their own initiative.

Several other persons may be concerned with control action taken in any one unit. Their activities may be affected indirectly by the control action, or the control data may be useful to them as background for future planning. For these purposes complete and reliable data become relatively more significant than promptness. So for these other persons, summarized data, accompanied by a report on action taken and a revised estimate of expected results, may be most useful.

Whatever the channels established, a significant aspect of a designed communication network is its dependability. Managers can proceed with confidence if they feel assured that significant information is flowing to them in a regular fashion.

Management by exception. A well-developed control system permits an executive to "manage by exception." In this technique an executive stipulates that if the operation is proceeding according to plan, or within a narrow range close to plan, he does not want to hear about it. Instead, he receives a report only when actual or predicted results deviate beyond the limit. He can then concentrate his attention on these exceptions. With well-balanced controls and clearly defined parts, such management-by-exception should be safe and it obviously simplifies the reporting process.

Evaluate and Take Corrective Action

At best, standards—measurements—predictions—only command attention. Until someone acts on the information and the signals generated, we have not achieved control. Evaluation and corrective action are essential concluding steps.

Confirm and elaborate the warnings. Many predictors used in a control system are not fully reliable. Thus, a jump in absenteeism, if caused by a flu epidemic, may be a false alarm with respect to morale; a surge in inquiries about services may reflect unusual publicity on a popular TV program rather than genuine interest of potential users. So before we act, someone should take a hard look at all the evidence bearing on the change in expected results.

In addition to confirming that previous expectations are no longer valid, the manager under pressure wants to know what *is* likely to happen if no modifications are made. How serious is the prospect? Cold weather that portends increased demand for fuel, for instance, may merely call for adjustments within the normal capacity of the distribution system; but the prospect of an acute fuel shortage, requiring the rationing of supplies and perhaps the closing of schools, theaters, and plants, triggers quite a different set of responses. Corrective action is based on a more thorough evaluation of what lies ahead.

Possible corrective actions. That popular control illustration, home-heating, is extremely simple. When the thermostat detects a need for a change, only one action is possible—turn the furnace on (or off if it is already running). Most managerial control situations are vastly more complicated (and hence not suitable for automation). The measurements are multi-dimensional, the opportunities are unique, and practical alternatives have not been predetermined. Instead, an array of new possible solutions must be devised.

The "cause" of a departure from normal will suggest alternatives in a stable operation. But in dynamic settings possible alternative actions range over shifts in resources, modified ways of operating, changes in incentives, new alliances—inside and outside the organizational unit. Incidentally, conceiving of good alternatives requires quite a different talent than objective measurement and predictions; some people are good at both but this is a rare combination.[4]

Acting on a selected alternative. The final choice of the corrective action to be used and its implementation should be the responsibility of a "line" executive or operator. Inevitably, the new thrust is intimately tied to the entire flow of activities, and we want to guard against creating other problems while resolving the particular matter that is temporarily the focus of attention. Control is sure to be ineffective unless a close tie is preserved between spotting the need for action and actually doing something about it.

Input for replanning. Although control mechanisms are primarily intended to keep the ship on its original course, they also supply useful information for reevaluating that course. Control will not always be entirely successful. After the corrective action is put into effect, a new forecast of expected results should be made. This

---

[4] Since corrective actions are so diverse, and so often feed back into earlier phases of management (planning, organizing, and leading), we do not examine them in this book. The interdependence between *control design*—which is our major concern—and other phases of management is discussed in Chapter 10.

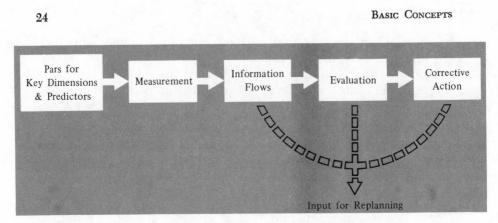

FIGURE 2-3. *Using Control Standards.*

forecast may continue to reveal some gap between expected and desired results. Hopefully the gap will have narrowed, but the magnitude of external changes, unexpected internal difficulties, or bad luck on recognized risks may still make our stated target unattainable. Or, with favorable breaks, the stated targets may now be too modest.

Some shift in plans is then called for. The easiest adjustment is to move the par without changing the key result factors. For instance, the date for completion of a program may be extended, or the number of successful conversions lowered. Nevertheless, one change often suggests another, and the new environment provides new opportunities. Replanning is needed. The data and revised forecast generated by the control activity then become part of the input for this replanning. A new cycle of planning and controlling is started.

Differences in Design Elements in the Three Control Types

Steering-controls have been the focus in the preceding review of elements that a manager manipulates when he designs a control system. Steering-controls are more sophisticated and have more elements than post-action and yes-no controls. So the elements involved in these two simpler types have already been discussed.

The elements in each control type are indicated in Table 2-1.

All managerial controls involve defining desired results, setting pars, and specifying the information flow. Post-action controls also include evaluation, but not corrective action. Yes-no controls may or may not lead to evaluation and corrective action; a "no" at some screening points simply kills the project or batch, but when the final

TABLE 2-1. *Differences in Design of Three Basic Types of Control*

Design Elements	Elements in the Design of Three Basic Types of Control		
	Steering-Controls	Yes-No Controls	Post-Action Controls
Define desired results	√	√	√
Look for predictors of results	√	—	—
Select composite feedbacks	√	—	—
Set "par" for each predictor and desired result	√	√	√
Specify information flow	√	√	√
Evaluate	√	sometimes	√
Take corrective action	√	sometimes	—

result is highly desirable an evaluation may show that re-work may overcome the reason for the "no."

Identifying predictors and combining several of these into composite feedbacks are not elements in yes-no or post-action controls. And these are the elements which usually offer the greatest potential for improving a control system.

The control elements reviewed in this chapter are derived from *company* goals and are cast in the *company* structure. The viewpoint and values in our discussion have centered on the *company*. An effective manager must think in these terms. However, the psychological response of the people who are affected by the controls must also be carefully weighed. This is the subject of the next chapter.

FOR FURTHER READING

CARROLL, S. J., and H. L. TOSI, *Management by Objectives*. New York: The Macmillan Company, 1973, Chapters 4, 5, and 6.
Discusses goal setting and performance review for individuals.

SAYLES, L. R., *Managerial Behavior*. New York: McGraw-Hill Book Company, 1964, Chapter 10.
A close look at the ways managers actually monitor operations, with suggestions for improving this phase of control.

SCHLEH, E. C., *Management by Results*. New York: McGraw-Hill Book Company, 1961, Chapter 14.
Sets control activities in basic, operational focus.

STOKES, P. M., *A Total Systems Approach to Management Control*. New York: American Management Associations, 1968.

chapter 3

HUMAN RESPONSES
TO CONTROL

A FIRE siren never put out a fire. Nor has an on-line computer printout secured a new customer. Only when some person responds to the signal or takes action in anticipation of it, does a managerial control become effective. An adjustment in behavior is crucial.[1]

For each control report we design, we should have a clear and comprehensive answer to the question: Who will modify his behavior because of this report, and in what way?

In fact, responses to controls may diverge widely from the purpose for which they were designed. The controls may be mistrusted and disregarded, and they may have significant side effects. So this chapter explores ways to generate positive responses to controls—and ways to minimize the negative reactions. Although people's feelings about controls vary widely, we do have some data on typical responses; behavioral scientists have described controlled behavior, and executives have reported on a wider range of experience. Our aim here is to translate these findings into guides that can be used in designing and operating a control system. These guides relate to:

1. Behavioral dimensions of the control elements singled out in Chapter 2.
2. Responses to the three basic types of control identified in Chapter 1.

1. Behavioral Dimensions of Control Elements

Each element in a control cycle can provoke constructive or negative responses. The goals that receive attention, the pars or standards set, the use of predictors, the reliability of measurements

[1] Automated response of a machine or physical system is an engineering achievement outside the scope of managerial control.

and reports, and the manner of corrective action—all affect the eagerness or sullenness of the people being controlled. So in addition to the rational, mission-focused aspects of control design discussed in the preceding chapter, we need to incorporate behavioral dimensions.

RELATE CONTROLS TO MEANINGFUL AND ACCEPTED GOALS OF THE PEOPLE WHOSE BEHAVIOR WE SEEK TO INFLUENCE

Meaningful goals. A desirable end result from the viewpoint of a central manager may be regarded as vague and other-worldly by an operating supervisor. The operating vice president of a large textile firm, for example, is deeply concerned that each company mill keeps its production costs in line with the quarterly budget; for him the financial budget provides the natural criteria for cost control. However, the mill foremen, who are in the best position to change costs, regard budgets as a nuisance. Most of the foremen realize the competitive necessity of keeping costs down, but in their eyes budgets merely absorb time that they could better devote to actually doing something about costs. The foremen think—and act— in terms of machine loading, output per man-hour, spoilage or material usage, machine maintenance to avoid stoppages, and indirect labor on the mill payroll. They know what happens to these factors long before budget reports are received, and explaining budget variances is merely a chore imposed upon them by "the pencil pushers in the office." Controls that have a constructive impact on the mill foremen must provide prompt data on operating factors.

"Client satisfaction," to cite another example, is a poor control criterion for the printing shop superintendent of a public relations firm. This particular man has unique talent for producing beautiful brochures, announcements, and reports. But he takes no part in deciding what message is important or what media are most suitable. His finest creations may, or may not, satisfy clients. Instead of the broader goal of client satisfaction, relevant control criteria for the printing superintendent are unique and attractive publications, on-time production, and reasonable costs.

A control criterion is meaningful to a person (1) when it is expressed in terms that are operational to him—that is, in terms of actions and results within his sphere of activities; (2) when he can significantly affect the outcome being considered; and (3) especially when the outcome is clearly measurable.

Accepted goals. To generate a constructive response, a control criterion must also be *accepted* as reflecting a valuable part of a job.

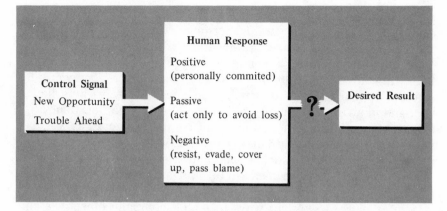

FIGURE 3-1. *Human Response—the "Intervening Variable" between Control Signals and Constructive Action.*

Psychologically, the person being influenced should feel that measuring his results in this particular respect is normal and legitimate.

"Why heckle me about late deliveries?" asked one irate shipping clerk, "I handle the orders as they come to me." Here, the control on late deliveries was probably causing more harm than good, at least for the individual who felt the pressure.

Goals may be accepted for a variety of reasons. The person accepting them feels that they are relevant to his job, they are the way the game is scored, they are worthy, they represent "professional" conduct, or they bring punishment or reward. Whatever its origin and reinforcement, psychological acceptance is a prerequisite for the success of any control. Without acceptance of the goal, the control is sure to be resented; evasion, manipulation of reports, buck-passing then become normal responses.

Acceptance is often passive. The person being controlled recognizes the objective as part of his responsibility, but beyond that he personally is indifferent to the outcome. The typical taxi driver, for example, conscientiously takes the passenger to the stated destination without the slightest concern with why the trip is being made. Many of us, in fact, have wide "zones of acceptance" with respect to parts of our work.

Although passive acceptance is adequate for control to function, active commitment to the result sought obviously is superior. Psychologists speak of a goal being "internalized"—the individual includes the aim as part of his own personal desires. When an individual gets a personal satisfaction from achieving a result which is also a company goal, his feeling about control shifts. Control becomes an aid to him in gaining personal satisfaction. Steering-controls, especially, become aids rather than irritants.

Active acceptance is common. Typically, the carpenter does take pride in the quality of his work; the teacher does want his students to learn; the gardener does like to see a flourishing flower bed; and likewise with many, many people. Although questions arise about levels of achievement and about competing goals, as we shall soon see, a completely indifferent person is very rare.

Control, then, is much easier and more effective when it is related to goals that are meaningful and actively accepted by the people who really shape the result. So when designing a control system, it pays to seek out ways to match company objectives and personal values.

Participation to secure understanding and acceptance. Participation in setting standards is widely advocated as a way of gaining positive commitment to control. Here is an example. Salesmen in a large frozen food company were asked to develop a picture of a first-class salesman in terms of his duties and his performance. Following a thorough and frank discussion of this ideal Mister X, each man was asked to prepare for himself a statement of what he thought he would accomplish during the next year. The men were expected to cover all the functions that had been listed for Mister X, but they were free to set whatever outputs they believed were reasonable.[2] Management then used these statements as standards of performance for the following year. The only adjustments—and these were made with the concurrence of the salesmen involved—were to scale down some of the outputs if the men had set too high a standard for themselves.

The success of participation in a wide variety of instances attests its usefulness. But the way participation is employed is critical. Hypocrisy is an ever-present danger. Usually the end result sought by a control is fixed by plans which are already settled. To pretend that these goals can be changed in participative discussions is misleading, and the participants will soon recognize this fact. Such discussions lead to a cynical mistrust of the whole system.

Participation does help, however, (1) to develop a mutual understanding of the aims and mechanisms used, (2) to translate

---

[2] No trickery was involved. The salesmen knew when they started discussing Mister X how the description would be used.

broad goals into criteria that are meaningful and operational for the persons being controlled, and (3) to set stimulating pars—as indicated in the following pages. These are the subjects on which the controllee can make definite contributions and having done so, is more likely to psychologically accept—and possibly feel a commitment to—the control endeavor.

SET TOUGH BUT ATTAINABLE PARS

Meaningful and acceptable control targets create a situation in which various control mechanisms can function. There is agreement about the aims of cooperative effort. However, further refinement of goals is necessary. The specific level of quality, amount of output, or degree of perfection which is expected also has to be agreed upon. So we turn now to the psychological aspects of establishing pars.

Much criticism of controls comes from trying to enforce "unreasonable" levels of achievement. Some kind of speed limit, sales quota, or deadline, for instance, may be quite acceptable; but tempers rise if the standard is felt to be impossible or unnecessary. Unacceptable pars turn positive effort to meet the target into all sorts of scheming to evade the pressure.

Dual purpose of pars. Pars serve two distinct purposes: (1) as a motivational target we hope to achieve, and (2) as an expected result used in planning and coordination. Although actual practice varies, most evidence indicates that people generally respond to a challenging target. We get more personal satisfaction and pride out of meeting a tough assignment than through exceeding an easy standard. Not everyone will meet the tough standard every time, but some will; and the overall result is higher than with an easy standard. Notice, however, that with such high pars some deficiencies will occur. For planning and coordination this slippage must be anticipated; thus, the estimated sales volume used for coordination purposes will be lower than the total sales quotas for individual salesmen or separate product groups.

Pars that motivate. Tough pars will motivate people only if several conditions are met. The individuals responding must feel that the target is attainable with reasonable effort and luck. Perhaps, like a handicap in golf, the person will privately set his aspirations a bit lower than the stated standard. But to generate determination

and willingness to be inconvenienced, he needs a personal belief
that he has a reasonable chance of success in achieving the adjusted
target.

Also, a supportive atmosphere is necessary. Supervisors and
staff can provide help; they *join in the game* of meeting a challenge
—like climbing a mountain or swimming the English Channel.[3]
Success is emphasized and rewarded; failure is a disappointment
but is not treated as a catastrophe. If the par can be adapted to
unpredictable, external variables—as with quotas tied to industry
activity or cost tied to orders processed—the feeling of being sup-
ported in the venture is increased.

Motivating pars cannot flaunt social norms. Peer groups have
their own ideas about acceptable behavior—output ceilings in a
factory is the classic example. If a control pushes a person to take
actions which are not approved by his friends, he is likely to abide
by their social standards. Of course, there are plenty of instances—
especially in the executive ranks—when social pressures support controls. The attitudes that really count are those of associates whose
friendship and respect a man wants to keep. If these persons feel
that a control standard and its measurement are fair and that cooperating with management is the right thing to do, they will con-
stitute a social force supporting that standard.

Between the two extremes of direct opposition and strong
support are many shades of group attitudes. Perhaps a group is
indifferent to what management wants to accomplish, but it may
have certain norms of its own, such as keeping the gang together
or deciding who may legitimately set a standard. So exactly how
peer groups affect responses to controls needs to be examined for
each case.

Pars that breed dissension.   If a par is so difficult that con-
trollees consider it "impossible" to achieve, a strong negative, emotional response is likely. In fact, the behavioral science literature is
so full of gruesome cases of unattainable pars that naive readers
assume that control always produces bad results. Sending incomplete or shoddy work to the next department, falsifying records,
and transferring blame are common devices used by workers under
pressure to appear to meet a standard. When such defenses are
inadequate, a person may become indifferent about his entire job,

---

[3] In *The Game of Budget Control* (London: Tavistock Publications Limited,
1968), G. H. Hofstede cites an array of psychological studies showing the role of
the game spirit in adult motivation.

irritable to work with, and hostile to his boss. To relieve his frustration he often joins in horseplay, slips a dead mouse into a can of
soup, and takes an active part in any protest movement that is
available.

Confronted with such behavior, a supervisor who wants to
meet his commitments often increases pressure on the alienated
operator. We are then faced with a vicious circle—more pressure,
more resistance.

One way to avoid such a collapse is to lower performance
standards to a level that the performer regards as realistic. Even if
this lower par is insufficient to attain some broader output or quality
objective, it is better than a standard that precipitates negative behavior of the kind just described. Fortunately, there are a variety
of other steps we can take to reconcile gaps between what is needed
and what the person responsible for the work regards as realistic,
such as redesigning the job, training, demonstrating, transferring
people, and the like. The point here is that ambitious standards
beyond a psychologically acceptable level can play havoc; the behavioral response may undermine the social system that the control
is intended to stimulate.

Participation in setting pars. Since feelings about what is
reasonable and what is not reasonable affect the response to a control so sharply, extra effort should be made to uncover those feelings. Participation in setting the pars provides this communication.
Each supervisor—from the president to the foreman—can frankly
discuss with his subordinates the levels of expected results that will
be used in each major control.

Such participation includes fact-finding, communication, prediction, negotiation, and mutual agreement. Although the supervisor
has the stronger bargaining position, sincere agreement by the subordinate is essential if the control is to induce a positive response.
And this feeling cannot be ordered by the boss. The process of
participation itself has beneficial side effects, but these depend on
agreeing upon standards that the subordinate really feels are attainable.[4]

Technical pars such as man-hours per telephone installed or
credit losses per dollar of sales tend to be stable and have to be
renegotiated only when significant changes occur in the environment

---

[4] The "Management by Objectives" technique, when properly applied, generates individualized objectives. These objectives become the acceptable pars for
control systems which we are discussing here.

or technology. Broader output pars like sales quotas or budgeted profits, on the other hand, are reset for each period of time.

Participation in setting the broader output pars is akin to bidding in the card game of contract bridge. A player first negotiates a tough but realistic standard based on his new situation and then strives to achieve the contract.

If our controls are to induce positive responses, then, we must approach the establishment of pars not just in terms of company needs. These standards also connote fairness, challenge, self-respect, social norms, winning, and related attributes for people. Consequently, the setting of tough but attainable pars calls for keen perception of the attitudes and values of the people whose behavior we hope to influence.

LIMIT NUMBER OF CONTROLS AND MINIMIZE
COMPETITION FOR ATTENTION

A third cluster of behavioral considerations, in addition to feelings about goals and about pars, relates to the total load. We must avoid "the straw that breaks the camel's back."

Every one of us is subject to a whole array of controls. This multitude of controls creates some psychological problems in addition to those already discussed. The combined total may be so oppressive that we rebel. Also, the various controls compete for attention and the stress of these conflicting pressures can lead to irrational, emotional response.

Consider the controls on a purchasing agent. Quality of materials and supplies obtained must meet exacting production standards. Delivery dates must anticipate actual use. Inventory levels will be checked against capital allocations. Prices paid will be measured in terms of cost estimates. Departmental operating expenses have to stay within budgets, and a wide variety of personnel and accounting procedures should be followed. No personal gifts can be accepted. In addition, there are informal controls on intangible factors such as obtaining data on new materials, responding to normal pressures for reciprocity, and minimizing risk arising from strikes and other shutdowns of suppliers. Tight controls over all these facets add up to a lot of pressure; the purchasing agent can justifiably feel that he is buffeted from all sides.

Psychological tolerance for controls. People differ in their desire for freedom—and in the particular areas where they feel that controls are repressive. One person may feel that regular working hours and scheduled tasks infringe on his rhythm of work, whereas another welcomes specific working assignments and checks on his progress but is irritated by controls designed to monitor how he gets the work done. To some extent, by carefully selecting people for specific jobs we can match these individual differences in security needs and in freedom needs with the number of controls inherent in the work assigned. However, in this age of reaction against "the establishment," the number of controls necessary for management are likely to seem excessive to most people.

An emphasis on steering-controls, rather than yes-no controls, will reduce the feelings of constraint. Although steering-controls may prod and signal a need for action, they do not restrict the action. Also, participation in selecting criteria and in setting pars —already recommended—helps to incorporate the resulting controls into the normal activities asociated with the job.

"Satisfactory" targets for minor criteria. The main way to make a variety of controls tolerable is to associate "satisfactory" levels of achievement with most of them. As long as a satisfactory level for, say, personnel turnover or equipment maintenance is achieved, no one gives it much attention; there is little or no pressure to improve performance beyond a satisfactory level. In any going concern, experienced personnel carry on many activities in this fashion; controls exist but most of them are rarely brought into play because people have learned to do satisfactory work. And these satisfied controls do not seem oppressive to the individuals operating under them.

Obviously, if only a satisfactory achievement is accepted for a particular criterion, additional improvement in that area will probably be sacrificed. In effect, we are saying that the potential benefit from a tighter control here is not worth the psychological cost and the reduced effort in other areas. So we must carefully select areas where the sacrifice of not pushing hard is relatively minor. Experience indicates that most people can give serious attention to only four to six different objectives. This rule-of-thumb suggests that controls above four to six in number should seek merely an adequate level of performance. Even four prime controls may be too many if they deal with complex and urgent matters.[5]

---

[5] A simple arrangement for dealing with targets that merely need to be "satisfied" is "Management by Exception"—a signal is raised only on the exceptional occasions when the satisfactory level of attainment is not being met.

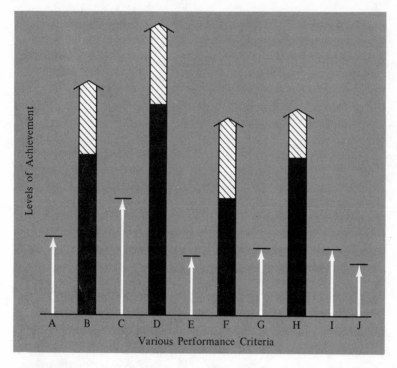

Pars for Multiple Goals

"Satisfactory" level for minor criteria

No-compromise level for major criteria

Desired level for major criteria

FIGURE 3-2. *Pars for Multiple Goals.*

Reduce competition for attention. Even a limited number of controls, each with a tough par, can place an operator in a psychological vise. For instance, the purchasing agent mentioned above may find pressures for ready availability of materials, high quality, low cost and low inventories competing for his attention. He may be forced to trade-off low cost for higher inventory, low cost for less quality, etc. He feels frustrated because he recognizes that meeting one goal will hurt him on some other front.

Such compound pressures can be relieved in several ways. Simple priorities may be established—in the preceding example instructions might be given the purchasing agent to meet targets in the following sequence: quality, availability, cost, inventory level. A more sophisticated guide would set "no compromise" levels below the desired pars, and then set a priority to fill the gaps between the "no compromise" levels and pars. In financial budgeting, three levels sometimes are specified for various accounts—optimistic, expected, and minimum. More often, an implicit tolerance range is understood by people using the controls.

The key in all these arrangements is to relieve the pressure at least to the extent of providing guidance for allocating effort among competing controls. By itself each control may be desirable and acceptable, but we must also consider how that control fits into the total.[6]

CONFINE DETAIL CONTROL TO SELF-ADJUSTMENT

Running his own show. Interference with the way a person does his work can be very annoying. A branch manager, for instance, may be fully committed to training new salesmen and have in operation his steps to meet an agreed-upon training target. But if he is then subjected to detailed control over the selection of trainees, their job assignments, and how they are supervised, he is likely to resent the control. Most experienced people—from bus drivers to atomic scientists—have a set of activities that they feel they know how to do well, and they regard interference by outsiders as lack of confidence and respect for this personal skill.

Expense budgets are a case in point. Such budgets frequently include a lot of detail. (The amount of detail often arises mainly from the availability of expense records.) Once specific items for, say, telephone or overtime are in the budget, a supervisor or staff controller is tempted to watch these items closely and insist on an explanation each month that actual expense exceeds the budget. The typical manager will resent such "needling," especially if his total expense is in line.

Feedback that assists self-adjustment. In behavioral terms, the person who feels that his domain is being invaded by a control is likely to be the one who can best initiate corrective action. He knows the local facts and is aware of the effect of manipulating one part of a total operation. Consequently, the control on over-

---

[6] Providing balance among an array of controls is examined more fully in Chapter 9.

time, for instance, will probably be most effective if the standard and the feedback on overtime become an integral part of *local* management.

In controlling detail, then, a desirable arrangement is to (1) design control mechanisms for elements worthy of systematic attention, but (2) route the feedback to the lowest level of decision-making for that element. The aim is to encourage self-control and to avoid interference by an outside person. Obviously such a scheme will work best when some evaluation of the overall results can be made, and when the local decision-maker recognizes that the detailed feedback information will aid him in achieving that desired overall result. In other words, for important details we design local, quick, operational feedbacks—and then encourage their use by the man who wants to run his own show.[7]

Conrad Hilton applied a variation of this arrangement in the management of his hotel chain. Each hotel, from the Waldorf-Astoria to the Shamrock, regularly computed and reported many ratios such as number of meals per guest-day and coffee shop sales per guest-day. Occasionally, at an unpredictable time (one manager claimed 1 A.M. to 3 A.M. was most likely), Mr. Hilton phoned a hotel manager to ask what was being done to correct an off-target situation. If corrective action was underway, Mr. Hilton was satisfied; he relied on the hotel manager to decide what action fit the local situation. The effect of the system was that local managers retained a feeling of autonomy in operating their hotels, but they were alert to the control mechanism that central management had designed for local use.

DEVELOP DISCERNING VIEW OF MEASUREMENTS

The human problems discussed thus far in this chapter relate to the design of a control system—what criteria are acceptable, how tough should standards be, how many controls can a person tolerate, how to build on desires for self-control. We now turn to a different kind of issue, the *credibility* of the system. Can we believe what the control reports say?

Your response to the gasoline gauge on your auto or to the

---

[7] This situation of considerable freedom in selecting local means to achieve an overall result is also well suited to participation in setting pars for local performance. Theoretically, the supervisor could withdraw entirely. In many cases, however, the supervisor wishes to strongly encourage the local operator to use particular controls, and periodic participation in setting pars is one way to indicate continuing respect for a control mechanism.

scales in your bathroom depends heavily on your belief in the accuracy and significance of the message the device is sending to you. Neither device is fully reliable, so you allow for a margin of error. But if the error is unexpectedly large, you become annoyed about the false alarm (or lack of alarm). Here, as with all controls, one's feeling about the measurement can affect the response significantly.

Attitude toward promptness vs. accuracy. As pointed out in Chapter 2, a prompt warning is often more useful than a tardy-precise one. For instance, flood warnings by the weather bureau fortunately turn out to be "wrong" (a flood does not occur) two-thirds of the time, but we don't want to wait until the water is at our doorstep before starting suitable action.

Psychologically, early but unreliable measurements are a potential source of tension. The person making the measurement (or prediction) may be criticized because of his "mistakes," and the person receiving the report may become resentful if he is pushed to act on "wrong" information. Only when all persons concerned recognize the inherent limitations of such measurements (or predictions) can such friction be avoided.[8]

Preliminary estimates and probabilities can be used for control in a variety of ways. Perhaps they merely alert the operator, or they may set in motion a series of more elaborate measurements. If a large number of similar events are involved—as in quality control of long runs of machine-made parts or in extending credit to customers of mail-order retailers—statistical ranges of normal variability help distinguish between random and significant deviations. In all such uses a clear awareness that the measurement itself may be misleading is coupled with precautionary action. Everyone knows that uncertainty exists. They are prepared to shrug off the false alarms.

Credibility attached to control data. Uncertainty arising from a small, early sample, as just discussed, is fairly easy to understand and to accept psychologically. A different sort of problem arises from the use of symptoms and of subjective measurements. Here the significance of the measurement is open to question. For instance, how valid a control instrument is the number of "laughs" at a Broadway play, or the number of lunch dates scheduled by an aspiring young management consultant?

[8] Some individuals are so reluctant to take risks that they are incapable of dealing with unreliable data. These risk-aversion people either cry "wolf" for every distant shadow they see or do nothing until they are sure the wolf is at the door. They are misfits in a dynamic control job.

Such information is often helpful feedback; it provides some additional "feel" about what is happening. But both the measuring and the evaluating can be challenged. And this doubt about its meaning makes such data poor input for strict controls.

"Soft" data can be used for *self*-control, and for supplementing more objective measurements. But until such a measurement has gained credibility in the minds of both the controller and the controllee, its use for yes-no controls or for post-action evaluation is likely to evoke a negative response.

2. Responses to Three Basic Types of Control

Behavioral reactions to each control element—such as those discussed in the preceding pages—are vital parts of every control design. To ensure that these human dimensions are recognized, we can also relate normal responses to each of the three basic types of control singled out in Chapter 1. The underlying response patterns are those already described, but regrouping them by types of control shows their significance in a new light.

POSITIVE RESPONSE TO STEERING-CONTROLS

The great virtue of steering-controls is that most people regard them as helpful rather than as pressure devices. If the goal is accepted, then the various feedbacks are treated as aids in achieving that desired result. Even granting that sometimes a control report conveys unwelcome news and prods a person to extra effort, the warning is constructive. Coming before work is completed, the signal is seen in terms of action needed rather than personal evaluation.

Since steering-controls provide inputs early enough for the principals concerned to use the data in their own decisions, their personal involvement in the control cycle is high. And this close involvement in the control process adds to the positive response.

Goal acceptance is crucial. Unless the astronaut wants to get to the moon, or the salesman wants to increase his sales, reports of deviation from course and of potential obstacles are merely so much static. And an unreasonable par can sour the reaction to the latest word about, say, competitors or planned shipping dates. Steering-controls generate a positive response only when the people on the giving and receiving ends of the control effort are steering in the same direction.

Who steers is also an issue. As already mentioned, outside regulation of detailed operations annoys persons who see themselves as experts in that area; consequently, self-regulation is more welcome. This suggests that steering-controls should be translated into action as close to actual operation as possible. The positive response to the control activity then spreads among operating personnel.

Too many reports can swamp the system. Unimportant information diverts attention from the main goals; frequent needling is irritating. So for many dimensions, "satisfactory" conditions should not be reported. Feedback should center on key variables and on major shifts in the work environment. Steering can then focus on goals we wish to maximize and on serious obstacles.

NEUTRAL OR NEGATIVE REACTION TO YES-NO CONTROLS

Yes-no controls set hurdles to be crossed. They ensure that quality standards are met, that a proposed action is within budgetary restraints, and the like. For the "professional" who takes pride in his work, being able to clear such hurdles easily may provide reassurance. But the check is only whether work is good enough to pass. If it is better than standard, little or no praise is given; if it is below standard, the work is rejected. And rejection of work often creates delays and resentment. On balance, when yes-no controls demand attention, they proclaim bad news.

Negative feelings about yes-no controls are often increased: (1) when a person is unable to achieve other goals because his work is blocked by this hurdle; (2) when the par is felt to be unreasonable (e.g., the budget is too tight or the requirement for a salary increase is too strict); or (3) when the standard is vague and unpredictable. The legality of a contract, the impact of a public relations release, or the qualifications of a person for promotion are typical examples in which standards are likely to be vague and unpredictable. Unpredictable standards are particularly troublesome because people lack guides on how to prepare for the control. Then if the standards applied to separate cases appear to be inconsistent, charges of favoritism and politics will follow.

Such reactions to yes-no controls can be reduced by, first, making clear that the control is necessary for attainment of company or department objectives. Both the aspect being measured and the par should be traceable directly to a basic objective. Second, keep the measurements as objective as possible and insist on consistency in their application.[9] These steps will rarely make the control popular but they will cut down frustration and foster a feeling of fair play.

POST-ACTION CONTROLS AS SCORE CARDS

In a strict sense, measurement and evaluation after work is completed cannot alter what is already done. Like Monday-morning quarterbacking, talking about what might have been won't change history. Nevertheless, as already indicated, post-action controls do serve two general purposes: (1) If we are going to play another game next week, the Monday morning review of successes and failures helps us *plan* the *next* engagement.[10] (2) If some kind of reward is tied to how well actual results match selected goals then the *anticipation* of that comparison and payoff may be a strong incentive.

The influence of anticipated rewards depends upon the strength of the rewards (or punishments) and upon the perceived basis on which the rewards will be allocated. We are not in this book exploring the nature of rewards—they vary from bonuses and promotions to commendations and scoring well in the game (or their opposites). But we are directly concerned with the score cards that determine, in fact, when a person receives a reward. Post-action control reports are such score cards.

Control design affects what is put down on that score card—the factors which are watched, how they are measured, and the expected levels of performance. Their impact on behavior has several dimensions.

1. People in the system will be sensitive to factors measured and reported; consequently, care in maintaining desired emphasis among objectives is important.

2. A lack of confidence in the reliability of measuring and re-

---

[9] Occasional exceptions will, of course, be necessary. In fact, people working under the system may strongly advocate exceptions—to achieve justice or meet an emergency. But exceptions have little meaning until we have established a stable, predictable social system as a base.

[10] Notice that analysis and evaluation for purposes of future planning need not stick to a control format of predetermined standards and feedback. The aim of the analysis is to help devise better plans, whereas control is predominantly concerned with restraining and motivating behavior toward selected objectives.

porting mechanisms will generate a feeling that granting of rewards is probably inequitable.

3. In a rapidly changing environment people may discover that their final score is affected more by their skill in renegotiating pars after-the-fact than by efforts to improve actual results.

When the purpose of a control is to produce a score card, several ways of automatically adjusting par after-the-fact are available. A "flexible budget" which is adjusted up or down on the basis of actual volume—sales quotas adjusted for actual disposable income in each territory, and "standard cost" adjusted for the actual product mix manufactured—illustrates an attempt to make the final standard reflect changes in the environment. Such devices usually increase the chances that the participants believe the par is fair—even though they recognize that the par may move up as well as down.

CONCLUSION

Managerial controls are concerned with achieving results—with a balance between inputs and outputs that pushes toward the company mission. These controls, however, take effect only when they influence the behavior of people. It is behavioral response, not the mechanics of a control, that really matters. So when designing a specific control or a control system, we must consider how executives and other people involved will react. This chapter highlights conclusions drawn from behavioral science studies that relate to the process of controlling.

Controls typically have a poor reputation, at least in terms of their popularity with persons being controlled. Fortunately, such a negative feeling need not prevail. By including behavioral aspects in the design and execution of controls, these devices can become normal aids in cooperative effort. Important in this respect are meaningful and accepted goals, challenging but attainable pars, restraint on the number of controls, means for resolving conflicting pressures, encouragement of self-adjustments, and acknowledged uncertainty in some of the measurements. Participation in designing and setting standards also helps.

These human considerations combined with the elements in all control cycles, outlined in the preceding chapter, are the building blocks for control design. Each successful control is a composite of these components. In the next five chapters we shall see how they can be applied to quite different types of situations.

FOR FURTHER READING

Numerous books report on behavioral science studies, but none covers the range of issues raised in this chapter. Moreover, they typically discuss matters not of direct relevance to a manager. Each work cited does contain some insights on the topic under which it is listed, though rarely is this its primary focus.

Alienated behavior

ARGYRIS, C., *Integrating the Individual and the Organization*. New York: John Wiley & Sons, 1964.

DALTON, M., *Men Who Manage*. New York: John Wiley & Sons, 1959.

DUBIN, R., *Human Relations in Administration*, 4th ed. Englewood Cliffs, N.J.: Prentice-Hall, Inc., 1974, Chapter 19.

Acceptance of goals, and legitimacy

KATZ, D., and R. L. KAHN, *The Social Psychology of Organizations*. New York: John Wiley & Sons, 1966.

LEAVITT, H. J., *Managerial Psychology*, 3rd ed. Chicago: University of Chicago Press, 1972.

Multiple goals

FILLEY, A. C., and R. J. HOUSE, *Managerial Processes and Organizational Behavior*. Englewood Cliffs, N.J.: Prentice-Hall, Inc, 1969.

RICHMAN, B. M., *Soviet Management*. Englewood Cliffs, N.J.: Prentice-Hall, Inc., 1965.

RIDGWAY, V. F., "Dysfunctional Consequences of Performance Measurements," in W. A. Hill and D. Egan, eds., *Readings in Organization Theory*. Boston: Allyn and Bacon, 1966, pp. 521–527.

Level of pars

HOFSTEDE, G. H., *The Game of Budget Control*. London: Tavistock Publications Limited, 1968, Chapters 3 and 8.

MILES, R. E., and R. C. VERGIN, "Behavioral Properties of Variance Controls," *California Management Review*, Spring 1966.

Participation

BLAU, P. M., and W. R. SCOTT, *Formal Organizations*. San Francisco: Chandler Publishing Company, 1962, Chapter 7.

HERZBERG, F., *Work and the Nature of Man*. New York: World Publishing Co., 1966.

LIKERT, R., *The Human Organization*. New York: McGraw-Hill Book Company, 1967.

TANNENBAUM, A. S., *Control in Organizations*. New York: McGraw-Hill Book Company, 1968.

II
ADAPTING DESIGN
TO DIFFERENT SITUATIONS

chapter 4

CONTROL OF
REPETITIVE OPERATIONS

MANAGERIAL controls deal with a wide array of activities. In this and the next four chapters we will look at several different types of situations calling for control action. The underlying thesis is that control instruments should be tailored to fit each situation. At the same time, the diverse controls which we select must be compatible with one another and merge into a viable total control structure.

Repetitive operations are the focus of this chapter. Following chapters deal with increasingly varied, long-run, and hard-to-measure activities.

Beginning with the Industrial Revolution, techniques for controlling repetitive operations have received close attention. Most of this work was done by engineers and focused on manufacturing processes. The resulting literature is voluminous and technical. Our purpose here is to distill from this rich experience a set of simple concepts that can be applied to repetitive operations far removed from the factory floor.

Most cooperative ventures are not, and should not be, patterned like a factory. On the other hand, virtually every human activity has within it repetitive elements. New undertakings—from lobster farming on the ocean floor to self-teaching clinics for high school drop-outs—always must develop and control an array of standard, dependable routines. By controlling these repetitive operations we provide a stable base and liberate energy for all sorts of dynamic activity.

The technical aspects of controlling repetitive operations are fully treated in handbooks which are readily available (see references at the end of the chapter). The following discussion, on the other hand, outlines in simple terms a series of basic steps applicable to all sorts of enterprises. Even relief missions and research laboratories should include control of their repetitive operations in their total control scheme.

Aim for Stability and Dependability

As in the rhythm and habits of our personal lives, repetitive activities provide an essential base upon which the more dramatic, unique actions are built. For any productive system—hospital, school, airline, or factory—to work well these repetitive operations must be dependable and predictable. Controls designed to assure this stability are vital and normal.

A pocket calculator, to cite a popular device, is worse than useless if it fails to work properly. Quality is essential. So in the production of thousands of identical units, standards are set for incoming materials, for processing, for sub-assemblies, and for finished products. Inspection—perhaps mechanized—is made at each stage, and prompt corrective action is taken whenever actual quality deviates beyond acceptable limits. Comparable quality control is applied to innumerable products, from contraceptives to tractors.

The use of resources for repetitive operations can also be carefully controlled. Standards are expressed in terms of man-hours, units of materials, machine hours, and/or dollars per unit of output. Then the common records of inputs and outputs provide an actual measurement which is compared with the standard ratio. These efficiency controls range from barrels of oil per kilowatt hour for electric utilities to man-days of work per income tax return audited by the Internal Revenue Service. Considerable refinement and sophistication may be necessary in the practical use of such ratios; nevertheless, they do contribute to the stability and dependability of the operations.

Single Out Repetitive Elements

Repetitive operations invite control for several reasons:

1. Considerable expense for setting standards and devising measuring schemes is warranted because the techniques developed can be applied to a large volume of operations.
2. Often a substantial bank of relevant historical experience and records is available for analysis in setting standards.

3. Opportunity exists for obtaining the views of persons being controlled and for them to learn to work with the control data.

These advantages may relate to only part of an operation. In newspaper publishing, for instance, the news and hence the specific content of a paper change daily; nevertheless, the typesetting, proof-reading, printing, distribution, and financial record-keeping are highly repetitive. Similarly, in hospitals patient care is quite individualistic whereas maintenance and office activities are stable. Such differences in activities suggest that one kind of control will be best for repetitive activities, whereas other forms of control will be applied to the varying elements. Clearly, however, controlling repetitive operations will be part of the total system.

Establish Normal Performance

Landing a jet plane at an airport requires that a host of features of the plane and of the airport be operating properly— landing gear, power, wing flaps, radio or radar, traffic control, the air strip, etc., etc. Extra finesse may be welcome but the critical issue is performance in a normal, expected manner. A variety of controls are exercised to assure that such normal conditions prevail. Only when conditions deviate from normal is special action necessary.

In every repetitive operation, a set of normal conditions, like those in an airplane landing, is necessary. So we set up controls on such matters as the flow of information, the training and preparation of people, the availability of supplies and equipment, the quality of work in previous stages of operation, and the rate and quality of current output. The approach laid out in Chapters 2 and 3 can be used as a guide in selecting factors and designing control mechanisms for a specific situation.

Most of these controls can operate on a management-by-exception basis—i.e., no action is taken unless the actual condition or a predictor of the condition is abnormal. For very critical items, however, yes-no controls should be introduced.

Implicit in the idea of normal operating conditions is a level of "satisfactory" performance. The controls are not geared to maximize quality, information flow, maintenance, and the like. Typically, incentives promote extra effort on only one or two features—say, customer satisfaction in the service department or number of placements in a public employment office. In general, however, the primary benefits arise from an established rhythm of work made possible by a controlled set of operating conditions and flow of information and supplies.[1]

Retain an Effective Balance

Existing records will make some aspects of a typical repetitive operation easy to control. Inputs of materials and labor are recorded for financial accounting; these costs can be related to output. Xerox machines already have counters, postage meters add relentlessly, customer orders generate sales slips, doctors' prescriptions tie in with medical records, and so on. Such ready-made data should, of course, be used whenever it is relevant.

Ease of measurement has undesirable side effects. We tend to overemphasize the characteristic—cost, quality, delivery—which is easy to measure, and to transfer pressures from those activities which are loosely controlled. This is an age-old problem: The Roman centurion who had to provide 1,000 soldiers on a specified date might scrimp on their training, which could not be easily detected. On the other hand, when quality is easily tested, as in the maneuverability of a ship, durability or cost may be sacrificed. In the current squabbles about safety of new drugs we know that tighter controls on safety slow up availability and increase costs.

So when designing controls for repetitive operations a whole cluster of characteristics must be considered, not just those features that are easy to measure. One of the major contributions made by Frederick Taylor and his scientific management theory was the insistence that all factors contributing to productivity be brought under control.[2] Taylor started with output per worker but soon was involved with the maintenance of machines and tools, the consistency of materials, the methods used, the availability of work to be done, the quality of output, and the training and instruction of workers. If a high standard of output per worker is to be set, all these contributing factors have to be brought under control. The total system has to be kept in balance.

[1] This concept of normal, acceptable levels of performance does not preclude technological progress. When a better way of performing the repetitive operation is found, a modified system with a new set of norms is established. After a learning period, the new system becomes the normal, acceptable one.

[2] The inadequate interpretation Taylor gave to worker motivation and supervisory practice should not be confused with his clear recognition that the entire shop had to be considered as a total system.

The need for balanced controls arises in a bank, a social security office, and a child day-care center, just as it does in a manufacturing plant. The services produced and the technology employed differ, but each has inputs of labor, materials, and equipment which are transformed into outputs with specified qualities and delivery times. Control of any one input or output will distort the overall operation unless other elements are also watched.

Simplify the Controls

Controlling is expensive. Setting standards, measuring, ·evaluating, reporting, and adjusting all involve effort and cost. Inevitably, controlling interferes to some extent with actions of people who are turning out the work. Moreover, as we saw in Chapter 3, a lot of control may stretch people's tolerance for regulation. Four approaches are available to reduce such costs in the control of repetitive operations.

1. *Sampling.* When a sample of a large volume can reasonably be expected to represent the total, only that small number need be measured. Sophisticated statistical techniques have been developed for applying this concept to quality control, and these techniques can be adapted to checking inventory, verifying deposit obligations of a bank, assessing public health, checking water pollution, and a variety of other situations.[3] If the sample is selected carefully, reliable control is obtained at a substantial reduction in cost.

2. *Acceptable deviation.* Closely related to sampling is the idea of acceptable deviation. In any human or mechanical process some variation from target is inevitable. Many of these deviations from standard are small and random. Unless we have to presume that they represent a persistent and perhaps increasing bias in results, we give them no further attention.

The practical question is how wide a variation in results can managers afford to disregard? Occasionally, for repetitive operations enough historical data exist to answer this question statistically.

---

[3] Statistical sampling was conceived for and is applied predominantly to repetitive operations in production. With ingenuity its use can be extended to many other situations being discussed in this chapter. For a clear statement of the statistical techniques involved see R. B. Fetter, *The Quality Control System.* Homewood, Illinois: Richard D. Irwin, Inc., 1967.

Usually, however, a subjective judgment is made about (1) the variability that is likely to occur even when the system is operating normally, and (2) the magnitude of variation which if permitted to continue will not have serious consequences. The smaller of these two numbers becomes the *acceptable deviation*. We then reduce control expense by not evaluating and undertaking corrective action if measured results fall within these limits.[4] For example, for many products a variation of monthly sales from budget of 2 percent is not worth investigating.

3. *Strategic control points.* Some points in an operation, like the observation towers used to locate forest fires, provide unusually good vantage points. For instance, a bunching of new orders, slow delivery of supplies, or high absenteeism may warn of impending trouble; or a final quality test may summarize an entire series of preceding operations. Careful watch at such strategic points may make regular review of intermediate points unnecessary; only if a warning or trouble shows up at the strategic points will a closer review be launched.[5]

4. *Delegation.* The chief way to maintain control of repetitive operations at a reasonable expense is to keep the bulk of the measuring and adjusting very close to the operation being controlled. Upper-level managers (and staff) should confine their attention to two things—designing effective control methods and then making sure that these methods are being used by operating people and first-line supervisors. The numerous adjustments which keep operations on an even keel are most easily made by the persons who do the work. So we make sure that they know the standards, that measurements and feedback data are readily available to them, and that help in correcting deviations is on call. With a workable system in place and understood, the remaining task for senior managers is limited to summary review and spot-checking to assure that that system is being used—a control on the control.

By some combination of these devices—sampling, acceptable

---

[4] A refinement on simple high and low permissible tolerance limits is to pay attention to persistent deviations in one direction even though they are small.

[5] Often a trade-off must be made between the probable savings arising from errors caught at a control point and the expense of catching those errors. However, when the "savings" are intangible the selection of control points rests on subjective judgment. For a fuller discussion of the selection of strategic control points, see W. H. Newman, *Administrative Action*, 2nd ed. Englewood Cliffs, N.J.: Prentice-Hall, Inc., 1963, p. 422–430.

deviations, strategic control points, and delegation—controls of repetitive operations should be simplified. Ideally, they should become normal and customary, just as the operations themselves. Although essential, the controls need not be complex and burdensome.

Cultivate Psychological Acceptance

Custom and habit naturally become important in repetitive work. Controls should also. By incorporating controls into normal behavior—like stopping one's car at a traffic light—feelings of repression can be reduced. Because performing repetitive operations may become boring and frustrating, care in the administration of their control takes on added significance.

An airline pilot is subjected to a host of restraints but he accepts them as necessary elements in his work and takes pride in a job well done. A similar attitude can be developed for the clerk handling social security payments, the telephone operator at a hospital, or the workers at an electric utility plant. Obviously, structuring and enlarging jobs so that the operator identifies his output with a recognizable accomplishment will assist in relating the task—routine though it may be—to an intrinsically valid objective. With such an attitude, controls are as natural a part of the job as a thermostat is in an air-conditioner.

The opportunities for participation in setting up controls of repetitive work are limited—changes are infrequent and the output typically must be fitted in with other activities. Nevertheless, management's openness to suggestions of alternative ways to achieve the same ends will contribute to a feeling of shared commitment.

The use of "satisfactory" norms—already recommended—relieves the sense of ever-increasing pressure, and explicit recognition of "acceptable ranges" in deviations cuts down on supervisory heckling. Even more helpful is the delegation of day-by-day adjustments; this contributes to an attitude of self-control instead of reliance on an outsider's standards of behavior.

Adapt to External Variables

Our attention thus far in this chapter has centered on achieving dependable, stable operations. By controlling the internal environment, repetitive tasks can be done in a predictable and economical manner. However, we know that this model is less and less suited to the dynamic external environment. Social, economic, political, and technological changes call for frequent adjustments, and control devices dealing with such changes are needed—as we shall see in the following chapters.

But external change does not make control of repetitive operations obsolete. Instead, we (1) create departments, sections, or even parts of jobs in which the benefits of stabilized activities can be retained, and (2) add buffers which absorb common types of change without upsetting the stabilized operation. Several control techniques are specifically designed for this second approach, notably standard cost accounting and flexible budgets.

Price changes, for instance, can distort cost standards. A person may use exactly the standard quantity of materials but find that his expenses are high (or low) because prices which are beyond his influence have gone up (or down). One way to remove this distortion is to use a standard price for internal record-keeping and charge any market deviations to a price variance account.

Likewise, volume of operations may change unexpectedly for any of a variety of reasons, and this creates question about what cost standards to use. A common way of keeping the standards fair is a prearranged adjustment in labor, materials, and other cost standards based on the degree of internal response a person is expected to make in such circumstances. Flexible budgets and similar devices can be used in this manner.

When several different products are made and the quantity of each varies independently, the adjustment of standards is more complicated. The typical standard cost accounting system meets this problem with an estimated labor and material cost per unit for each product; then after the end of a period the quantity of each product actually processed is multiplied by its unit standard. The resulting standard allowance for labor for all products is totaled and this total is compared with actual payrolls. A composite standard for materials is computed in the same manner.

In all such schemes the aim is to make adjustments for the external variable in such a way that the integrity of the internal system is maintained. The levels of expected achievement, the pars, are adjusted or the effect of the external variable is screened out. These revisions, it is hoped, will maintain the fairness and the significance of a comparison between the control standard and the measure of performance.

FIGURE 4-1

Flexible Budget for Housekeeping Labor Costs in a Motel.

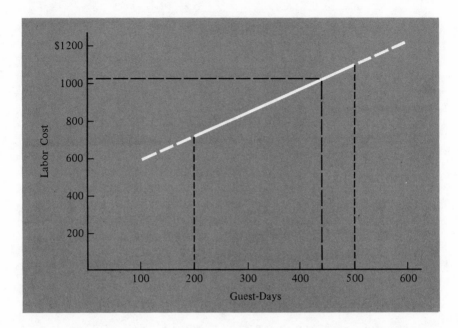

Post-Action Control Standard That Can Be Adjusted for Actual Volume.

	Low Volume	*High Volume*
Guest-days	200	500
Labor costs	$690	$1095
Average increase per guest-day ($1095 − $690 ÷ (500 − 200)		$1.35
Fixed costs $690 − (200 × $1.35), or $1095 − (500 × $1.35)		$420

Budgeted cost if volume is 275 guest-days
$420 + (275 × $1.35) = $791.25

Budgeted cost if volume is 437 guest-days
$420 + (437 × $1.35) = $1009.95

Conclusion

Whenever we begin to design a set of controls for, say, a store, fire department, or Buddhist monastery we can be sure that many of the controls will deal with repetitive activities. For intangible services, just as for the production of steel, a stable and predictable base of regular operations is needed. Dynamic leaps take off from such a base.

The exhaustive work already done in the field of production suggests an approach applicable to almost any repetitive work. The underlying concepts are: Single out repetitive elements, establish normal performance for these elements, retain a balance among the controls, simplify the controls and promote self-control, cultivate psychological acceptance, and build in adaptability to external variables. Experience indicates that many diverse measurements and feedbacks are needed to keep high quality work flowing promptly and efficiently. But most of these controls can be localized and be made a normal part of the total behavioral pattern. The aim is dependability and predictability for the more routine operations.

FOR FURTHER READING

GREENE, J. H., *Production and Inventory Control Handbook*. New York: McGraw-Hill Book Company, 1970.
Like other production handbooks, provides many specific suggestions for simplifying, standardizing, and then controlling production operations.

JURAN, J. M., ed., *Quality Control Handbook*, 2nd ed. New York: McGraw-Hill Book Company, 1962.
Comprehensive handbook on quality control edited by a leading authority in the field.

MUNDEL, M. E., *Motion and Time Study: Principles and Practice*, 4th ed. Englewood Cliffs, N.J.: Prentice-Hall, Inc., 1970.
Authoritative treatment of simplifying repetitive operations, and setting output standards for them.

SHILLINGLAW, G., *Cost Accounting: Analysis and Control*, 3rd ed. Homewood, Illinois: Richard D. Irwin, Inc., 1972.
Full treatment of cost accounting from managerial viewpoint. See

Chapter 15 for standard costing for control of labor and material costs.

STARR, M. K., *Production Management Systems and Synthesis.* Englewood Cliffs, N.J.: Prentice-Hall, Inc., 1972.
A more readable exposition of modern quantitative treatment of production problems, including a good chapter on quality control.

chapter 5

PROJECT AND
PROGRAM CONTROL

CONTROL of programs and projects poses special problems. Both programs and projects involve a series of interdependent steps to be performed on a stipulated schedule. A project, however, is narrower in scope than a program; it has a clear-cut mission and a sharp terminal point when that mission is completed. Our analysis will be simpler by looking first at project control.

The aim of project control is to assure that (1) the mission is accomplished (2) on time (3) with the resource allocated. Thus, if a manager undertakes to air-condition the office, hire four additional salesmen, or raise money for a new hospital wing, he is concerned with *when* the project will be done and what it will *cost* in terms of human effort, materials, and capital—as well as with the specific objective. Since the objective is clear, few control problems arise regarding the level and measurement of the primary result. Timing and use of resources are more difficult to keep in view.

Keeping on Schedule

After a project is completed, knowing that it was a year late serves little purpose. Instead, we need timing control that aids completion on the target date. The basic approach is simple: (1) Divide the total project into steps. (2) Note any necessary sequences among the steps (e.g., legal approval before public announcement of a bond issue). (3) Decide who will be accountable for each step. (4) Determine resources needed and their availability for each step. (5) Estimate the elapsed time required to complete each step. (6) Assign definite dates for the beginning and ending of each step, based on (2), (4), and (5).[1]

[1] The estimates of resources and of time required may be high or low. The accuracy of these predictions will depend partly on the stable base of repetitive

FIGURE 5-1. *Gantt Chart Showing Job Assignments for Far East Facility Study.*

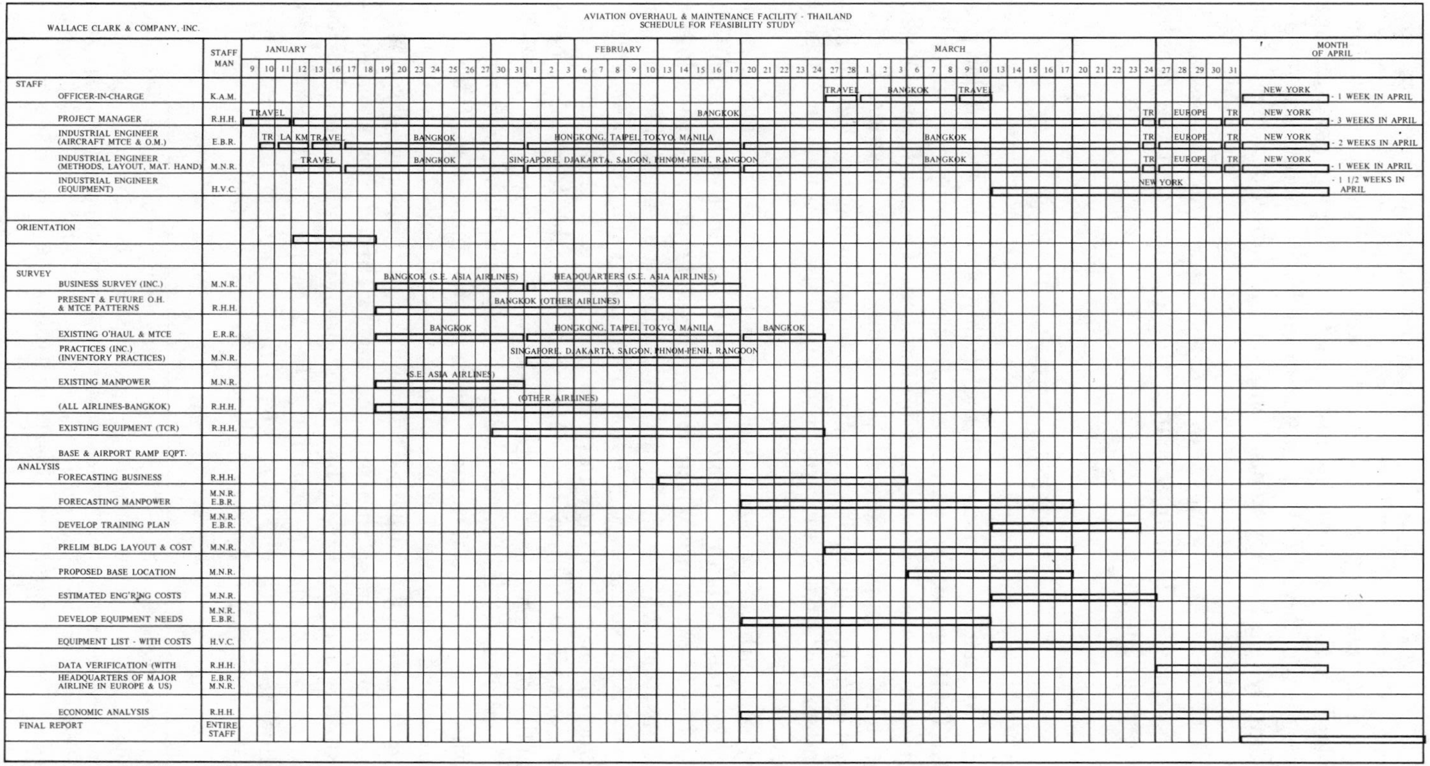
WALLACE CLARK & COMPANY, INC.
AVIATION OVERHAUL & MAINTENANCE FACILITY - THAILAND
SCHEDULE FOR FEASIBILITY STUDY
STAFF MAN
JANUARY
FEBRUARY
MARCH
MONTH OF APRIL
9 10 11 12 13 16 17 18 19 20 23 24 25 26 27 30 31 1 2 3 6 7 8 9 10 13 14 15 16 17 20 21 22 23 24 27 28 1 2 3 6 7 8 9 10 13 14 15 16 17 20 21 22 23 24 27 28 29 30 31
STAFF
OFFICER-IN-CHARGE
K.A.M.
TRAVEL
BANGKOK
TRAVEL
NEW YORK
- 1 WEEK IN APRIL
PROJECT MANAGER
R.H.H.
TRAVEL
BANGKOK
TR
EUROPE
TR
NEW YORK
- 3 WEEKS IN APRIL
INDUSTRIAL ENGINEER (AIRCRAFT MTCE & O.M.)
E.B.R.
TR
LA
KM
TRAVEL
BANGKOK
HONGKONG, TAIPEI, TOKYO, MANILA
BANGKOK
TR
EUROPE
TR
NEW YORK
- 2 WEEKS IN APRIL
INDUSTRIAL ENGINEER (METHODS, LAYOUT, MAT. HAND)
M.N.R.
TRAVEL
BANGKOK
SINGAPORE, DJAKARTA, SAIGON, PHNOM-PENH, RANGOON
BANGKOK
TR
EUROPE
TR
NEW YORK
- 1 WEEK IN APRIL
INDUSTRIAL ENGINEER (EQUIPMENT)
H.V.C.
NEW YORK
- 1 1/2 WEEKS IN APRIL
ORIENTATION
SURVEY
BUSINESS SURVEY (INC.)
M.N.R.
BANGKOK (S.E. ASIA AIRLINES)
HEADQUARTERS (S.E. ASIA AIRLINES)
PRESENT & FUTURE O.H. & MTCE PATTERNS
R.H.H.
BANGKOK (OTHER AIRLINES)
EXISTING O'HAUL & MTCE
E.R.R.
BANGKOK
HONGKONG, TAIPEI, TOKYO, MANILA
BANGKOK
PRACTICES (INC.) (INVENTORY PRACTICES)
M.N.R.
SINGAPORE, DJAKARTA, SAIGON, PHNOM-PENH, RANGOON
EXISTING MANPOWER
M.N.R.
(S.E. ASIA AIRLINES)
(ALL AIRLINES-BANGKOK)
R.H.H.
(OTHER AIRLINES)
EXISTING EQUIPMENT (TCR)
R.H.H.
BASE & AIRPORT RAMP EQPT.
ANALYSIS
FORECASTING BUSINESS
R.H.H.
FORECASTING MANPOWER
M.N.R. E.B.R.
DEVELOP TRAINING PLAN
M.N.R. E.B.R.
PRELIM BLDG LAYOUT & COST
M.N.R.
PROPOSED BASE LOCATION
M.N.R.
ESTIMATED ENG'R'NG COSTS
M.N.R.
DEVELOP EQUIPMENT NEEDS
M.N.R. E.B.R.
EQUIPMENT LIST - WITH COSTS
H.V.C.
DATA VERIFICATION (WITH
R.H.H.
HEADQUARTERS OF MAJOR AIRLINE IN EUROPE & US)
E.B.R. M.N.R.
ECONOMIC ANALYSIS
R.H.H.
FINAL REPORT
ENTIRE STAFF

Courtesy of Wallace Clark & Company Incorporated.

The resulting schedule is the key to virtually all timing controls. It is the standard against which actual progress is compared. By spotting deviations in any step corrective action can be begun promptly, and the potential impact on subsequent steps can be quickly traced.

Gantt charts, which can range in size from a sheet of paper to an entire wall, help visualize the schedule and the progress to date (see Figure 5-1).

PERT for Complex Projects

To analyze and control the timing aspects of a major project, PERT (Program Evaluation and Review Technique) explicitly lays out even more detail than Gantt charts. Originally developed for the highly complex task of producing Polaris missiles, PERT has been adapted to a wide variety of undertakings, including hospital fund-raising drives and construction of the World Trade Center in New York City.

As with any control, we start with a plan of action. Suppose our goal is to launch a new product or place a communications satellite in orbit and we arrive at the stage at which we know the actions that will be necessary to achieve our goal. The first phase of a PERT analysis is to note carefully each of these steps, the sequence in which they must be performed, and the time required for each. This information is recorded in the form of a network, usually on a chart such as those shown in Figures 5-2 and 5-3.

The chart in Figure 5-2 is highly simplified so that we can easily grasp the main features. It shows the main steps that a United States auto equipment manufacturer would have to follow to market a new anti-smoke muffler. The manufacturer has purchased a tested European patent so that engineering to United States requirements is simple, and he already has a well-organized plant and distribution setup. Arrows on the chart indicate the sequence of events he must follow to get his new product on the market, and the numbers on the arrows show the required time for each step.

---

operations—the control of which is discussed in the preceding chapter. As in military planning, the dependable behavior of operating units based on s.o.p.'s and thorough training serves as the building blocks for the design of campaigns. Insofar as the programs call for new and different activities, the accuracy of the control standards drops. So control of repetitive operations aids project and program control.

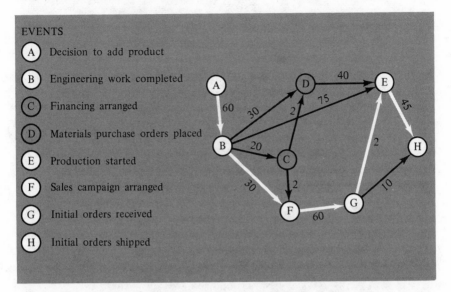

EVENTS

(A) Decision to add product

(B) Engineering work completed

(C) Financing arranged

(D) Materials purchase orders placed

(E) Production started

(F) Sales campaign arranged

(G) Initial orders received

(H) Initial orders shipped

FIGURE 5-2. *A Simplified PERT Chart. Events—that is, the start or completion of a step—are indicated by circles. Arrows show the sequence between events. The time (in days) required to move from one event to another appears on each arrow. The critical path—the longest sequence—is shown in gray.*

The chief advantage of expressing a program as a network is its emphasis on sequences and interrelationships.

PERT is typically applied to a much more complicated network than the illustration we have just used. Even this new product example was over-simplified; in practice, each of the major steps would be programmed in more detail. The preparation of the sales campaign, for instance, would involve packaging, pricing, sales brochures, installation manuals, training of salesmen, placing of advertisements, and the like; and each of these activities should be shown separately in the network. Such delineation improves our chances of catching delays early, and it also spells out the need for coordination at numerous points.

For a complex project, like the construction of a large plant, the network becomes complicated indeed. One network with 137 events is shown in Figure 5-3. PERT is especially suited to large "single-use" programs having clear-cut steps and measurable output.

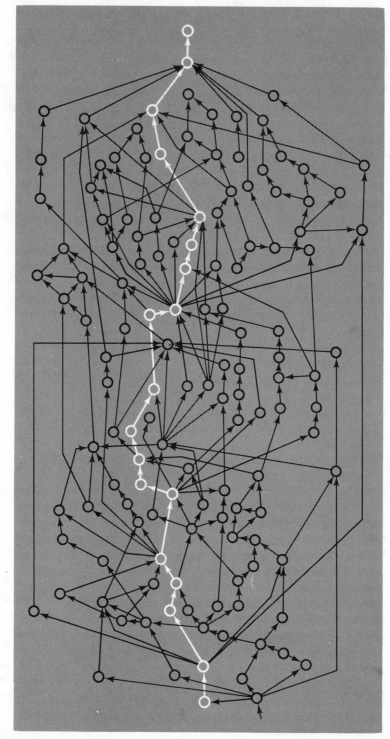

FIGURE 5-3. *PERT in an Actual Situation.*

The Critical Path

Because we are focusing here on time, we wish to know where delay will be most serious. A PERT network is very helpful for this purpose. By tracing each necessary sequence and adding the time estimates for each step, we can identify which sequence will require the maximum time. This is the "critical path." Other sequences will take less time and hence are less critical.

The critical path is especially important in planning and control. Any delay along this path will postpone the completion date of the entire project. On the other hand, knowing in advance which series of steps are critical, we may be able to re-plan (allocate more resources, perform part of the work simultaneously, and so on) in order to shorten the total time. In other words, (1) we focus control where it is most essential, (2) we are in a good position to spot potential trouble early, and (3) we can avoid putting pressure on activities that will not speed up final completion.

Moreover, as work progresses, reports on activities that are ahead or behind schedule will enable us to reexamine the timing. Perhaps an unexpected delay has created a new critical path. (For example, if tooling up—B to E in Figure 5-2—required an additional thirty days, a new critical path would be created.) Then corrective action can be shifted to the new sequence where no slack time exists.

Financial Accounting and Project Control

Gantt charts, PERT, and critical path analysis deal with timing —when things happen. Additional controls are needed to keep track of costs—the resources used to make things happen.

Financial accounting helps. *If* those outlays [2] directly associated with a project are recorded in a separate account, the total outlays can be matched with corresponding estimates prepared when the project was approved. Such "post-action" control will provide some guidance in estimating costs of similar projects in the future. And by rewarding persons who succeed in completing projects within estimated outlays, an expectation is created about the kind of behavior that gets rewarded.

---

[2] The outlay may be treated as an expense *or* an investment. This is an accounting issue, and is not significant in controlling the project per se.

Typical financial accounting unfortunately has serious limitations for project control. (1) Usually outlays are not closely matched with progress while a project is underway. (2) Outlays are measured in dollars—one step removed from man-hours and materials which are the realities that project managers actually manipulate. Unless a way to overcome these limitations is injected, financial accounting provides only post-action control.

Step-by-Step Resource Usage

To keep resource usage within bounds we need to tie outlays to rate of progress. The steps within a project provide a convenient way to measure progress, and the person accountable for each step should be known (see no's. (1) and (3) at the bottom of p. 60). So, the trick is to estimate and subsequently record outlays for each step.

Such step-by-step observation is the essence of PERT/Cost. A cost is estimated for each step in the PERT diagram, and then the actual cost for each step is compared with this standard. Also from these figures an accumulated estimated and actual cost to date can be easily computed at any stage of the project. Of course, although a PERT diagram is convenient, it is not essential; similar step-by-step estimates and measurements can be made for simpler projects provided each step is clearly identified.

The value of these interim figures lies in the opportunity to take corrective action promptly before the project is completed. The early steps will have become history but time will still be available to adjust (even abandon) the latter steps. For instance, an airplane manufacturer set up a subassembly plant in a depressed urban center, partly to provide local employment. Controls revealed that the early training steps were much more costly than estimated, so the nature of work was modified and the timetable extended. Without step-by-step standards and measurements, the committed administrators of this project would have been reluctant to make revisions when they did.

Step-by-step resource control may be expressed only in dollar figures. Often, however, standards and measurements will have more meaning if they are expressed in man-days or pounds of material. These standards relate directly to getting the project done. For example, if Bob and three assistants are allowed an extra week to complete a demonstration model, we know resource usage is creeping up and we know why.

To cite another concrete example, in an automobile subassembly plant tests revealed the need to use a stronger brake cable to fulfill safety requirements; this change not only increased costs, but also affected both the specific items to purchase and the men and tools needed for assembly operations. Fortunately, the project control system kept track of specific facts—not just dollars—and warned purchasing and assembly supervisors of the necessary adjustment.

How detailed should our project control system be? The dilemma is between losing touch with reality versus having many diverse standards and reports. We try to resolve this dilemma by (1) guessing when deviations from plan may have serious repercussions and concentrating control on those situations; (2) creating the mechanisms for control but then have them used primarily by operators and first-line supervisors—i.e., decentralizing the control activity; and/or (3) through selection and training, establishing professional behavior which includes self-imposed controls.

The control system will be simpler if we move strongly toward (2) decentralization and (3) professionalization. Insofar as we do, however, higher-level management will have to rely on voluntary reports of unusual situations and on occasional checking of how the decentralized system is working. If timing of performance is carefully monitored, wide deviations from schedule may also give warning of unexpected use of resources.

Summarizing: While informal supervisory control is adequate for very small projects, as soon as a project involves several departments and extends over several months a more systematic method of control is needed. Gantt charts, PERT networks, and critical path analysis focus on completing a project on time. Step-by-step cost control such as PERT/Cost helps keep dollar cost in line, but fully effective resource control has to be expressed in physical units. Such controls move us into even more elaborate observations and reports, and soon the burden of the control system is greater than its benefit. Fortunately we can decentralize much of this detailed control and encourage professionalization, which enables us to introduce control devices and concepts into the overall management design without adding greatly to the tasks of senior management.

Differences in Project Control and Program Control

The mechanisms for control of the (1) timing and (2) resource usage of a project can readily be adapted to programs. Typically, program control will deal with larger and larger aggregates of resources. If closer control is desired, the program can be divided into projects and these projects controlled in the manner just outlined.

Consider two examples. A mid-continent oil company decided to significantly increase its crude oil supply from off-shore wells in the Gulf of Mexico. A program involving the following steps was needed: purchase of oil rights leases, exploratory drilling, development drilling, construction of collection systems, ties to pipelines and refineries, and—for each move—financing, government permissions, public relations, joint ventures, deals with contractors, manpower, and organization. Clearly a well-controlled program became crucial since a major blunder could wreck the company. The control problem was similar to that faced by a Boston commercial bank which decided to increase its involvement in black neighborhoods. This involved opening branch offices prepared to make auto and personal loans, cash checks, attract savings and checking accounts, and the like; in addition, loans to black businesses and on local real estate had to be expanded; and clearly additional black employees (including officers) had to be selected and trained and arrangements made for transferring personnel between the new branches and headquarters. Here again, the program could be controlled by dividing it into a series of steps—some of which could be treated as clear-cut projects.

Control of programs, in contrast to project control, is more likely to involve revisions of the objectives and/or the pars. Programs require more flexibility in their administration because (1) the planning premises—official assumptions regarding external conditions—upon which the programs are based may not be valid, (2) unexpected internal success or obstacles may arise, or (3) details of future steps or their timing may deliberately have been left vague because of uncertainties.

Monitoring Program Premises

Every program is based on premises—assumptions (official predictions) regarding the need for action, availability of resources, cooperation and approvals forthcoming, and similar environmental conditions. If these premises change, the program will lose some of its relevance or feasibility. Strikes, new legislation, price shifts, wars, and discoveries illustrate external events that may significantly affect a program.

Steering-controls especially require current reassessment of planning premises. So we need mechanisms to monitor the key assumptions. Such monitoring involves, first, arrangements to keep well informed of events and announced plans in the environmental sector. Second, predictions covering the period of the program are needed. Someone should be designated to assemble these data and forecasts and to communicate them to the people directing and controlling the program. The current information may be helpful in evaluating progress to date, and the forecasts affect the desirability of continuing on the original course.

Unless this monitoring is effectively done, aggressive executives will be using premises that suit their ambitions and passive executives will be pursuing programs which are no longer desirable.

Milestones in Program Control

Since many programs continue over an extended period, senior executives need an occasional opportunity for review during the execution phases. PERT and PERT/Cost contain too much detail, and for some general programs such specific control is not feasible.

"Milestones" provide a key for when to make interim reviews. Typically, as a program progresses points are reached when large commitments of resources must be made—for example, a new plant built, a large number of people employed, a publicity campaign started, or a registration statement for new bonds filed with the S.E.C. Such points—just before the next big bite is taken—are natural occasions for careful review. Another type of milestone is when a body of key information becomes available—say, a market test has just been completed, a legal opinion or decision obtained, laboratory tests completed, or the like. These new facts remove previous uncertainties or modify predictions.

At important milestones the entire program should be reexamined. Current appraisal of the need (mission) to be served, progress to date, resources already used and commitments made, competition, unanticipated difficulties, future need and availability of resources, chances for success, reassessment of future steps in the program and their timing—all come up for review. The milestone is made the occasion for an examination of the total program. There is no stigma attached to such a review or presumption that the program is in trouble; rather, the review is a normal, prearranged, prudent control.

Milestones do *not* match the calendar year; instead, they occur at times related to the progress of the work. Consequently, milestone checks supplement or replace annual budget reviews.

Acceptance of Project and Program Control

Psychological acceptance of project and program control typically poses no serious problem. Normally, supervisors who execute a project or program also actively participate in its preparation. They help prepare estimates of time and resources required for their respective parts of the program. And they are aware of the need for adjustments to fit all the parts together to achieve a well-defined goal. So, a feeling that unreasonable standards are imposed by an outsider is less likely to arise than in repetitive work.

Moreover, since each project and program is unique in some respects, occasional deviations from the estimates are expected. The emphasis is on steering-control rather than post-action appraisal, and the operators are involved in devising adjustments that will keep the venture on target. Of course, pressure for results exists and a particular standard may be regarded as too tough. However, the conflict between oppressed workers and over-zealous managers, so popular in behavioral science literature, is uncommon in project and program control.

Lateral tensions do arise. The steps in a project or program are often interdependent. A delay in an early step upsets the schedule and creates the need to speed up later steps. Likewise, low quality work at one step—perhaps resulting from an effort to meet time or expense standards—may add to the work at a subsequent step. A person struggling to meet a delivery deadline or a publication date does not think kindly of someone in a previous operation who let three weeks slip by through an oversight. Unavoidably, one man pays for another man's sins. Project and program control does not create these interactions but it may highlight them. Fortunately, the reason for the pressures can be made quite clear and neither the control mechanisms nor a capricious manager is blamed.

Much more serious from the viewpoint of control design are the conflicts that often arise between persons committed to completing a project or program and equally committed persons whose viewpoint is the protection and allocation of resources. We turn to this issue in the next chapter.

Control in Small Enterprises

The main concepts reviewed in this chapter apply to both large and small enterprises. Large ventures provide dramatic examples, but similar issues face managers of architectural firms, retirement communities, and even the modern dairy farmer. All have projects to control—and the same six steps in laying out a schedule, Gantt charts or their equivalent for checking progress, critical path analysis, and step-by-step resources control fit small projects as well as large ones. Elaborateness differs; underlying concepts remain the same.

Likewise with program control. Monitoring the program premises and full measurement at milestones are excellent techniques regardless of the magnitude of resources involved. Since fewer persons are involved, the formality and the paper work can be sharply reduced in smaller undertakings. However, a conscious stepping back for evaluation at key stages is sound practice for all programs.

The disturbing fact is that in both large and small enterprises control of programs is often haphazard. Once started, we tend to pursue a program until something obviously goes wrong. We neglect using steering-controls systematically, probably because we become so engrossed in the challenging tasks. Instead, we should decide in advance when and how each major program is to be evaluated. Doing so increases the chances of spotting opportunities for improvements as well as catching potential trouble early.

FOR FURTHER READING

CLELAND, D. I., and W. R. KING, eds., *Systems, Organizations, Analysis, Management.* New York: McGraw-Hill Book Company, 1969.
Includes readings on project management, program budgeting, and scheduling in defense plants.

MOORE, F. G., and R. JABLONSKI, *Production Control,* 3rd ed. New York: McGraw-Hill Book Company, 1969, Chapters 20 and 21.
Techniques of production scheduling, many of which are applicable to general programming.

NEWMAN, W. H., C. E. SUMMER, and E. K. WARREN, *The Process of Management,* 3rd ed. Englewood Cliffs, N.J.: Prentice-Hall, Inc., 1972, Chapter 19.
Expands on programming steps listed on page 60 and discusses adaptive long-range programs.

SCHODERBEK, P. P., ed., *Management Systems,* 2nd ed. New York: John Wiley & Sons, 1971.
Part III includes excellent articles on PERT, PERT/Cost, and related techniques. The total book is more concerned with systems than with programming.

chapter 6

CONTROL OF RESOURCES

CONTROLLING *activities*—repetitive activities and then projects and programs—and the *results* of these activities was our focus in the two preceding chapters. Another approach is to think in terms of the *resources* used in these activities. By controlling resources— the things, the people, the capital—we can regulate the activities and indirectly the results, and at the same time we seek to conserve the resources.

Dual Aim of Resource Control

In designing resource controls, these two broad objectives are often intertwined. One deals with stewardship of the assets—for example, preservation of capital, plant maintenance, conservation of human resources. Of course the assets are to be used, but this is done reluctantly and with an eye to replenishment. The second control objective is to use asset control as a mechanism for regulating diverse activities. Thus, if there is tight control on electric power or office space, the available supply is given to those activities we want to encourage and denied to activities we wish to restrain. The aim here is not concerned with conservation of the resource, although that may also occur; rather, the allocation is a device for permitting only certain kinds of work to proceed. Failure to distinguish between conservation and regulation in control design and administration invites serious conflicts.

Three kinds of resource control warrant consideration in designing most control systems—control of (1) physical, (2) human, and (3) financial resources. In particular situations other resources require attention, but the array of issues and potential devices will be illustrated in a brief review of these three familiar resource categories. We will move from relatively clear-cut physical resources to increasingly intangible resources.

Each type of resource has its own unique control problems. Although we shall deal with these, our analysis will keep returning to the more fundamental issue in control design: to what extent should regulation of the flow of resources be employed as a double check on operations?

Physical Asset Control

MAINTENANCE AND USE OF EQUIPMENT

Prudent managers take good care of their equipment. Care of an automobile is a familiar example. We have in our minds, if not on paper, *standards* of how our auto should function, and as soon as the engine sputters or we hear a warning squeak we have the source of the trouble fixed. Many of us go even further: (1) we have the auto *inspected* periodically to look for trouble, and (2) we do *preventive maintenance*—oil changes, replacing of filters, etc.—to avoid trouble. Sometimes this maintenance work interferes with our normal use of the auto, and it entails expense; so, there is a recurring question of what, when, and how much maintenance to undertake.

Similar issues arise for every kind of equipment and buildings, from typewriters to TV broadcast stations. In a major overhaul of an oil refinery hundreds of thousands of dollars are involved in the cost of repair and interruption of service; PERT is used to control the overhaul itself, and preventive maintenance is based on statistical probabilities. The doctor, farmer, recreation director, research scientist, as well as the production manager, all face nagging questions of how far to go in setting equipment standards, inspecting, repairing, and doing preventive maintenance.

As soon as a dozen or more people are working in the same venture, the tasks of equipment maintenance are likely to be separated from the central activities and assigned to a specialized unit.[1] This separation aids control and has other advantages: (1) Adequate attention to keeping facilities in good shape is assured. (2) Other personnel are relieved of a diverting task and can concentrate on their major missions. (3) Additional skill and knowledge can be developed in the specialized unit. Notice, however, that while

---

[1] Smaller firms often achieve the same result through a "service contract" with the equipment supplier, just like any one of us may turn the maintenance of our car over to a local garage.

maintenance work is gaining in attention and skill, the separation complicates coordination. For instance, when should operations be interrupted to avoid a possible breakdown in the future? Should a screening signal be ignored in a busy season even though this increases the risk of major trouble? Will users become careless with equipment because they don't have to fix it themselves? The manager of the total activity discovers that the improved control over maintenance is not clear gain. A point is reached at which better maintenance is not worth its indirect costs.

This balancing problem continues if we decide to expand the maintenance function. For example, in addition to maintenance we may decide to charge a "building manager" with keeping costs of running the building low, and with preparing recommendations for building changes or expansion if he thinks these will be needed in the future. He now exercises stewardship for a total facility and is expected to protect, maintain, expand or adjust if necessary, and assure economical use. He provides a necessary resource—much like the landlord of a midtown office building. By creating this stewardship arrangement a manager obtains a means for tighter control over the building.

We know that the stronger and tighter we control one aspect of an operation, the more necessary are balancing checks. So when we create stewardship over a facility such as a building, the need for counterbalance should be weighed. Specifically, the *quality* of the service provided has to be watched; the steward might become miserly. Fortunately, the users of the service will be concerned with and can easily observe its quality, so the additional elements needed to control quality are (1) agreed-upon standards for the service to be provided (key characteristics and pars), and (2) a communication channel for reporting deviations to individuals who can and will take corrective action.

Related to quality is who gets served first. Rarely are facilities large enough to supply all users at all times, so questions of priority

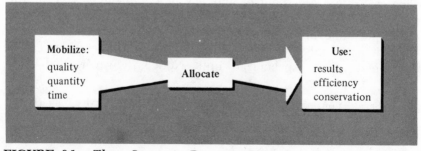

FIGURE 6-1. *Three Stages in Resource Control.*

or allocation arise. Access to a central computer or first call on the Xerox machine, for instance, often become troublesome questions. Unless some system of allocation has been agreed upon, internal politics or personal friendship with the custodian of the facility will determine who gets good service.

The need for a service can be used deliberately as a basis of control. A classic case is John D. Rockefeller's domination of the oil industry based upon his control of railroad tank cars. In countries where telephones are in short supply, the allocation of new phones among and within companies determines where expansion takes place. In a department store, the allocation of selling space can shape the merchandising strategy, and likewise for shelf space in a supermarket. Usually, allocation of the use of physical facilities is not the most convenient bottleneck control, but its potential exists.

Notice how the scope of resource control can expand. We start with simple maintenance control to assure availability of the needed resource. Then, to improve this control and for economy, a separate unit of organization is created. Next, the steward of the resource looks after future development and efficiency in the use of the resource: protection from abuse now ranks along with contribution to final output. Finally, control over the allocation and manner of use of the resource may become a means of regulating the basic operations and through them the final results.

A similar potential expansion in the scope of control exists for all other critical resources, as we shall see in the following pages. The design issue is how far to go along this path for each resource.

CONTROL OF PHYSICAL INVENTORY

An array of control issues and possibilities similar to those related to fixed assets surround inventory. The primary concern is assuring a supply of just what is required, in adequate quantity when and where needed, at a reasonable expense. A modest control system focuses on the stockroom where quality checks, physical storage, systematic issuance, and notification of need to reorder takes place.[2] As with maintenance, these activities are often separated to secure adequate attention and develop expert skill and knowledge. But as soon as this is done, the process becomes increasingly standardized and is likely to become less responsive to unusual operating needs.

[2] See Chapters 3, 4, and 5 for guides regarding the specific forms that control on such matters might take.

A much broader scope is created if we combine purchasing and stock-keeping. Then the entire function of providing one kind of resource is centralized. Materials control is facilitated, and the risk that people in this service activity will try to impose their ideas on operations is also increased. For instance, the influence that a purchasing agent should exert on the quality of materials actually supplied to operations is a perennial source of debate. If materials supply is a strategic factor for success in our industry, we may elect to create such a cross-check on operations. Or when handling a dangerous item like radium or drugs in a hospital, surveillance over its use may be warranted. But we should be aware of the additional complexity we are introducing in our control design.

In unusual circumstances the allocation of materials can be used as a fundamental control mechanism. In World War II, for instance, the primary technique for regulating the entire industrial effort of the nation—from the production of alarm clocks to airplanes —was the allocation of three critical materials—steel, copper, and aluminum. However, such extended applications of materials control are rare and usually confined to periods of temporary materials shortages.

SERVICE DEPARTMENTS VIEWED AS RESOURCES

Service departments generally can be viewed as resources. For example, the X-ray clinic in a hospital, the engineering division in an auto plant, the art department in an advertising agency, or the public relations office in a university each provide a valuable service to their respective enterprises. Each is a resource needed by other departments.

Control of such resources can be patterned on our approach to equipment and inventory control. The service activity is set apart so that it can provide concentrated attention to its mission and can be more easily controlled. Control focuses on maintaining the desired capability, on the quality of services actually provided, and on keeping expenses in line.

Separate services, however, always create a dilemma. The very reason for their existence is to get dedicated professional commitment to the specialized service they provide. But this commitment differs from that of the operating people who are predominantly concerned with converting resources into end-products. Although certainly not opposed, nevertheless their viewpoints, values, objectives, and criteria do differ. The tighter we control either the operating or the service people, the more likely are such differences to turn into severe stress. Consequently, when we place controls on the quality and cost of separate services, we should at the same time provide means for assuring that the services are adapted to the specific needs of their users.

Summarizing: Since physical assets and service units provide valuable inputs to operating activities, steering-controls to assure continuing capability are indeed prudent. Likewise, control over outlays—expenses and/or investment—should be included in our total design. Physical asset and service resources often are important enough to warrant separate, professionalized attention; but with such independence comes a need to assure responsiveness to operating requirements. In our zeal to "manage" a resource well, we must remember that the justification for having the resource is its contribution to final results. Rarely do we have controls that measure the delivery of contribution on a continuing basis. Instead, we rely on the users of the service to press for needed inputs. Consequently, a balance between the controls on operations and on the resource becomes critical.

Human Resource Control

The people who contribute their efforts to an enterprise are another of its major resources. Every wise manager is concerned about maintaining and enhancing this resource, and about using it effectively. So any broad design for managerial control must include arrangements for keeping track of the reservoir of human talent.

Control of human resources poses very difficult measurement problems. (1) The resource is not a stable element. The abilities and attitudes of people are ever-changing; and unlike equipment, people's behavior is influenced by the objective sought. (2) Reasons for changes in the output of people are difficult to pin down. In the complex chain of causes and effects, we are not sure of where to attempt to measure and control. As Figure 6-2 suggests, we can observe activities which are intended to improve our human resources, but tracing the effects of such activities through to improved output in operating situations is hard.

The general problem of selecting where in a process to exercise control, already discussed on pages 15 to 19, is illustrated in Figure 6-2, which deals with personnel training. We can focus on the training activity itself— A in the diagram. Or we can try to measure the direct results of that activity— B in the diagram. However, since the purpose of the training is to improve competence of

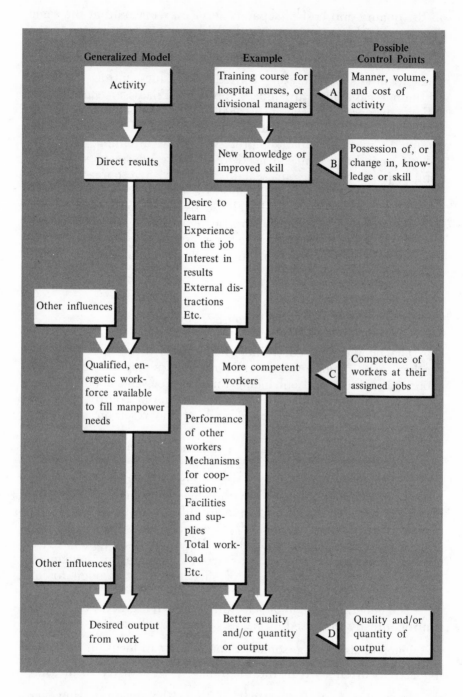

FIGURE 6-2. *Measuring Results of Personnel Activity.*

Control point △A focuses on the process. In the example, the way the course was conducted, the number of persons completing it, and its direct and perhaps indirect costs are measured. Control of such matters obviously is of interest to the first-level supervisor of the activity, say the training director. But it is of limited value to line managers.

Control point △B checks on the direct results of the training activity. Measurement may prove difficult because we are interested in the application of the new knowledge or skill to operating situations, not just apparent improved capability immediately following the course. Procedural acts or superficial use of new lingo are easier to observe than more subtle and often more significant changes in behavior, and consequently controls may emphasize the former.

Control at point △C focuses on total capability. The training input is intended to deal with only part of the total competence, but competence is also affected by other inputs. Balance, synergy, and side effects are involved. The line manager is concerned with the total direct and indirect effects. Unfortunately, the total capability is hard to measure—we often use only symptoms or non-representative incidents. And even when measured the cause of an observed change is often hard to trace back to a specific personnel activity, especially after the lapse of several months or years.

Control at point △D is desirable because it represents the operational goal for which the service was created. But so many other influences are at work that the contribution of the staff activity is hard to trace and isolate.

workers, we can focus on changes in competence—△C in the diagram. But competence is a means of improving output, so even more relevant to company objectives is to measure changes in output —△D in the diagram. Obviously, the dilemma is that as our control moves closer to the results we seek, the more difficult it becomes to identify the impact of the training.

A variety of techniques for evaluating personnel activities have been devised. Although these are often narrow in scope and not very reliable, they can be combined so as to provide some control on the supply side of human resources. (The effective *use* of human resources is so entwined with other operating and managerial activities that it rarely yields to separate evaluation.) Human resources are so crucial that a closer look at these potential control measurements is desirable.

For maintenance and development of human resources—the supply side—our control design may include some combination of the following: [3]

[3] "Human resources accounting" is still in such an experimental stage it can provide little help in actual managerial control. A variety of devices for evaluating specific personnel activities, such as a safety program or sensitivity training, are also omitted from the broad purview.

1. Individual appraisals   ⚠A

2. Checks on efforts to assure fair treatment   ⚠B

3. Assessment of union relations   ⚠B

4. Reports on filling specific manpower needs   ⚠C

5. General personnel indexes   ⚠C

The symbols relate these techniques to the hierarchy of controls shown in Figure 6-2.

INDIVIDUAL APPRAISALS

A whole library of studies on how to evaluate and counsel individuals is available. In recent years the conclusions strongly point away from rating a person's abilities; instead, the emphasis is on meeting performance goals which are periodically revised to reflect past achievements and future needs. The *performance* evaluation is more objective. It is directly related to what a man does; and fresh data are available for successive interviews—with individuals who have reached their peak as well as those likely to be promoted. If a firm is stressing "management by objectives," operating results will be reviewed in that program, and the individual evaluations will merely summarize the facts related to the particular person for the past period.

The performance evaluations normally give an individual excellent feedback on "How am I doing?" and coaching on how to improve or prepare for promotion follows naturally when appropriate. We have here a first-class man-to-man control device.

Individual performance reviews, however, do not generate data that can be summarized for the company's manpower base as a whole. The old-fashioned rating schemes did give scores which often wound up in neat statistical tables. But the basic ratings were ambiguous and unreliable; supervisors tended to inflate the evaluations to reward subordinates and to make their sections look good. Consequently, with rare exceptions individual appraisal plans do not provide a useful overall measurement.

Individual performance reviews, then, merit a place in the design of human resource controls, but they are inherently very localized. The most senior management can do is to train people in the technique and then establish controls to ensure that the reviews occur. The quality of the reviews, however, depends largely upon the two personalities involved in each review.

CHECKS ON EFFORTS TO ASSURE FAIR TREATMENT

The development of a good reservoir of manpower is aided by assurance of fair treatment. The negative side is even stronger; the *absence* of equitable wages, open employment, promotion on merit, or benefit programs at least comparable to those of other employers is very likely to result in a poor supply of labor.

Controls in this area can focus on the immediate results of policies and activities designed to assure fair treatment. For example, annual regional salary surveys compared to actual company pay rates will show whether company pay is in line with community and industry rates; the number of blacks in different job categories will show the progress a company is making on integrating its work force; hospital and health benefits, pension provisions, and similar perquisites can be compared with those provided by competing employers; and so forth. Unfortunately, the manner in which such arrangements are administered and the way employees (and potential employees) feel about them is more difficult to measure. At least, controls can be devised which should inform central management if the company is vulnerable on "equity" or "fairness" grounds.

ASSESSMENT OF UNION RELATIONS

Labor unions, originally concentrated in production and mining operations, are now found among farm, hospital, city, bank, and many other non-industrial workers. These unions significantly affect the availability and conditions under which human resources can be obtained. So a good control design in any company that has a union should provide advance warning of what the union expects and its likely tactics. This kind of information differs sharply from the objective data discussed in the preceding paragraphs, yet it is no less crucial.

Indicators are available, such as the openness of communications and the disposition of grievances through the established procedures. But unions are presumed to be protagonists, and tradition requires elected union officers to exaggerate their demands. Within the union members differ in their wants and often feelings are not well articulated. So forecasts for use in steering-controls are typically subjective impressions made by people with high empathy for union members and with a sensitivity to union politics.

Few companies have formalized the gathering and evaluation of this kind of information. We call attention to the need here partly to illustrate the intangible character of parts of every *comprehensive* control structure.

QUALIFIED CANDIDATES FOR VACANCIES

One obvious test of an adequate supply of manpower is the prompt availability of qualified candidates to fill vacancies. Information on this matter can be garnered quite easily from supervisors who wish to fill the vacancies. And if it is company policy to promote from within, the source of the candidates is simple to check. Such facts do not reveal how well qualified the successful candidates are nor even that the best available persons were selected. But they do provide a gross indication of the adequacy of a company's reservoir of human resources.

The measures just suggested are really post-action controls. Steering-controls would, of course, be preferable. For this purpose we need (1) manning tables which forecast the need for new appointees due to promotions, retirements, resignations, and expansion; and (2) lists of potential candidates for each specific opening who will be qualified by the time each job is expected to open up. The reliability of data of this nature is almost impossible to check; both the forecast of needs and prediction of the future ability of candidates are highly subjective judgments. So our control design probably will go no further than merely verifying that advance manpower planning is being done.

GENERAL PERSONNEL INDEXES

None of the measures discussed—individual appraisals, checks on fair treatment and on union relations, and reports on filling vacancies—gives adequate warning that a firm's human resources are deteriorating, nor do they provide a broad assessment. Some kind of a sensitive general personnel index is needed.

A number of symptoms of personnel health can be monitored. Several of these are significant for the specific condition they reflect, but they also indicate indirectly how well various personnel activities are meeting current needs. *Changes* in any of the following are at least suggestive:

Absentee rates	Number of grievances filed
Turnover rates	Wage level relative to the industry
Labor costs per unit of output	Ratio of direct and indirect payroll
Frequency and severity of accidents	Ratio of overtime to straight time
	Work stoppages

No one of these criteria is satisfactory because each may move up or down (or fail to do so) for unusual reasons such as a flu epidemic or the opening of a new branch. To overcome such false leads, we can prepare a composite index based on eight or ten factors, or we can show a "profile" of changes graphically.

For enterprises with fairly stable operations and, say, *at least* ten years' experience, a general personnel index serves as a useful barometer. The firm's relations with its manpower resources probably cannot change significantly—for the better or the worse—without a corresponding change in the index. The drawbacks of such an index are implicit in its construction; it doesn't point to the reasons for a change; it has short lead-time, at least in terms of when we would hope to initiate corrective action; and it is not suited to sharply changing situations in the environment or in the firm. Nevertheless, until a better measure is invented we probably will want some such measuring device in our overall control design.[4]

YES-NO SCREENING OF RECOMMENDED APPOINTMENTS

Adequacy of manpower supply is the question probed by the controls already discussed. A very different kind of control is screening proposals for new appointments—both new employees and candidates for promotions. This is a yes-no check—should the specific appointment be made?

Normally, yes-no review of recommended appointments is done by line executives. At least two supervisory levels should agree, because favoritism is always possible and the evaluation of a

---

[4] Independently administered morale surveys have a high potential. However, they must be skillfully made on a recurring basis, and this proves to be very expensive in terms of the outside help and the inside time consumed.

person's future potential is largely subjective. The control design question is whether an outside expert should *also* have veto power, and on what standards should the judgment be based.

Personnel departments start as a specialized service, providing (along with other help) assistance in recruiting, testing, periodic appraisal, and personnel record-keeping. Along with such service naturally come recommendations of whom to appoint to fill specific posts. But should the personnel advisor also be a controller? The answer is "yes," in a restricted way. The bases on which the personnel specialist is permitted a veto should be clear and limited. Among the legitimate criteria for his veto are meeting legal and policy standards on discrimination, consideration of potential candidates throughout the company, sincere comparison of qualifications with available candidates outside the company, consideration of total relevant experience and test scores, and the like. These are procedural matters intended to make sure that such selection is carefully weighed in light of the best evidence available. Rarely should we grant the personnel expert power to veto because he questions the future capability of a candidate. Typically, the personnel expert should advise but not control. The reasons for this constraint are (1) the judgment is predominantly subjective, (2) the line executives are the ones who are most deeply concerned with working proficiency over an extended period, and (3) the personnel expert will be more effective in his total role if he maintains the posture of an advisor.[5]

The screening control just described serves a useful though modest purpose. It assures that several desirable procedures have been followed. However, it does not deal with the flow of good

---

[5] Government civil service practice differs. Examinations (which include credit for experience) are centrally administered. Operating executives must then select from the top three or four persons on the resulting rosters. Here, primary reliance for selection is placed on the examinations so as to avoid any possibility of political favoritism. Unfortunately, the examination process is only partially reliable and is expensive; it assumes that the best prospects will take the examinations. For nongovernmental enterprises the risks of favoritism are not great enough to offset these serious handicaps.

Moreover, so-called "line budgets" which require a separate line in the annual budget for each person on the payroll are widely used in government (and universities). No new appointments or salary increases beyond the budget are permitted without approval of the authorizing body. A more stultifying straitjacket is hard to imagine. The emphasis is on dollar expenses rather than quality of personnel, and rarely is there a tie-in between the budget lines and desired results. All these devices are designed to hold a rein on change and expansion by checking inputs of human resources.

candidates into the screening process nor their effective use afterward.

Conclusion. The controls available in the human resource area deal primarily with (1) personnel activities—what activity took place, how much did it cost, what were its direct results; (2) comparisons of objective data on wages, fringe benefits, accidents, and the like, with other companies; and (3) assuring that desirable procedures for making personnel selections are followed. Unfortunately, these devices are far short of what is desirable in this vital area. Better steering-controls on the supply of specific talents and improved measurement of manpower utilization should be devised.

Control of Financial Resources

Financial controls are the best developed, pervasive tools for regulating business activities that we have. Every manager, even in non-profit enterprises, is familiar with the irrepressible, concise, financial reports which command his attention—whether the news be good or bad. Our purpose here is to place these tools in the context of a total control design and to point out ways to make financial mechanisms more effective parts of a balanced control system.

The underlying objectives pursued with financial controls are multiple, and this intermixing of objectives is a prime source of trouble in administering such controls. Two chief purposes, akin to those applied to other resources, are (1) preservation of capital resources, and (2) promoting efficient use of capital. In addition, (3) achieving corporate financial goals—an outcome of all operations—is layered on top.

Strains arise even among these three strictly financial purposes. For example, minimizing the risk of capital loss may interfere with actions that probably will increase the return on equity. More troublesome are short-run conflicts between financial controls and other controls. Consequently, care is essential in the way financial controls are designed and administered.

These issues will be revealed in a brief review of (1) capital investment control, (2) current asset control, (3) expense control, and (4) financial budgets. Notice that these controls deal almost entirely with the allocation and use of financial resources—not with supply.

CAPITAL INVESTMENT CONTROL

Yes-no control on capital investment proposals are widely used. Typically, a brief verbal description of each project is accompanied by estimates of total dollar outlays and of predicted reduction in expenses and/or increases in revenue. The expected benefits are then related to the outlays in an estimated rate of return or payback period. Comparison of the resulting ratios (or discounted rate of return) against a standard becomes the primary basis for approving or rejecting the proposal. Our concern here is not with the highly sophisticated techniques which can be used in making these calculations. Instead, the purpose of the control and the behavioral responses are of chief interest.

Approval or rejection of a request for capital is a powerful instrument. Perhaps for this reason, manipulation of data and of standards is all too common. Approval of capital investment projects is necessarily based on estimates, the reliability of which will not be known for several years. Often the persons making the estimates will have moved on to other jobs before results are apparent. In the meantime, unanticipated changes, inside and outside the firm, will occur. And the effect of a single decision is difficult to isolate. Although actual disbursements are normally tied directly to the amount approved for investment, only rarely is a subsequent attempt made to compare actual results with estimates of benefits made in the original proposal.

Under these circumstances the temptation is great to pad the estimate. The people reviewing the estimates are aware of the temptation, and some of them routinely scale down the benefits (or set required rates of return which have the same effect) on the assumption that the estimates are too high. So the estimators inflate the figures because they anticipate the scaling down, and soon we find people playing games with the data and the standards.

Several steps are desirable to at least reduce this tendency to maneuver the controls. First, to avoid arbitrary treatment, executives should distinguish among the purposes of the various proposals.

Cost reduction proposals are straightforward; company products (or service) and volume are left unchanged, and the estimated economies can be related to the necessary investment.

Many outlays, however, are for replacement or to meet new requirements (e.g., pollution abatement), and no saving is expected.

Such *necessary* projects cannot be weighed in terms of return on investment.[6] Judgments must be based on urgency and the required qualitative level of improvement.

Expansion of capacity requires estimates of when the added capacity will be utilized and whether all expenses and incomes will rise proportionately. Since uncertainty is greater, the ratio of expected return should be higher than for cost reductions.

Product improvements may be defensive or offensive. Defensive improvements have to be treated like "necessary" projects, whereas offensive improvements are usually tied to expansion, with the uncertainties already suggested.

Strategic investments such as new product lines or vertical integration entail even more uncertainties and consequently should have a higher prospective return.

The standards used for screening projects clearly should be adjusted to their purpose, as the above list suggests. Variations in uncertainty, urgency, and availability of other critical factors such as manpower all deserve attention. Also, balance is desirable; that is, one category of projects should not always take priority over another. Obviously, more than strictly financial inputs should be incorporated in the yes-no decisions.

A second way to reduce playing games with capital expenditures proposals is occasionally to check actual benefits against the estimates—and to impose penalties on people who made gross errors. The aim is to keep the estimating "honest" (like a judge occasionally handing down a tough sentence to warn would-be violators). This approach has serious limitations. The results of a single outlay are very difficult to isolate over a period of years; no check is possible on proposals that were rejected; and hindsight on risky ventures has little meaning.[7] So we should resort to this kind of a

---

[6] Discontinuance of the entire undertaking can be analyzed, but if the venture is to continue the necessary "tools" must be provided. Of course, if alternative ways of accomplishing the tasks are available, then an investment/benefit analysis of the differences between the alternatives is appropriate. Incidentally, cost reduction projects fall in this latter category; they compare a new method with continuing the old method.

[7] The use of probability estimates, which improves the basis for decision, complicates comparisons of projected results with actual results. If we really mean that there is an uncontrollable 20 percent chance of failure, and the actual result is failure, the responsible executive may legitimately claim that his work unfortunately fell within the 20 percent. Of course, if we identify in advance the source of the risk—say, a war or a new invention—and later find out whether that set of events occurred, then control is improved because we have shifted from an abstract statistic to an explicit planning premise.

check on estimates only when the figures appear deliberately inflated, and the estimator should be warned that a check will be made.

Turning a capital expenditure proposal into a project, and subjecting it to project control as described in Chapter 5, is a third approach. In effect, this replaces the single yes-no decision with a series of reviews at milestones. Financial people should, of course, participate in these reviews. Although well suited to very large proposals, such project control is too expensive for most proposals being screened.

Fourth, we can view capital expenditure control as part of the overall responsibility of each major operating executive. This presumes that he is accountable for total results, including return on investment and risks of loss. Since he now makes the yes-no decisions, he would merely be fooling himself to inflate estimates. Financial people continue to participate in reviewing proposals but under this approach they act in an advisory capacity.

To maintain a balance among controls and to reduce playing games with estimates, the fourth approach is desirable. It places financial people in a role like that suggested for personnel advisors on key appointments. They analyze from their perspective and recommend, but they don't veto.[8] By shifting their role to one of assistance and collaboration, their constructive contribution will probably increase.

CURRENT ASSET CONTROL

Accounts receivable and inventory are usually the main current assets into which capital flows. Controls focus not only on the aggregate amounts tied up; the liquidity as indicated by age, turnover, and resale value, and the likely write-offs are watched as well. For each of these aspects, standards based on past experience and industry norms are geared to sales volume and company financial structure. Then monthly or quarterly analysis of actual receivables and inventory provide the basis for fairly tight post-action control.

But who sets the standards? The controls just listed are strongly slanted toward conserving and protecting capital. These financial measures give little attention to the benefits gained from

---

[8] Many variations are possible. Sometimes if disagreement exists on larger proposals, they are appealed to a higher level of line supervision. Alternatively, the staff man is under obligation to notify his boss and the higher line man when he disagrees with a decision. Then, as a by-product of which decisions are upheld or reversed, a "common law" or informal policy develops.

such assets—the additional customers who deal with the company because of credit made available to them, and the improvements in service and production made possible by the inventories.

Production planning and control techniques can provide a good basis for regulating *physical* inventories; such controls do balance the financial pressures for conservation. In retail establishments merchandising programs create "open to buy" standards; these provide a marketing viewpoint on inventories needed. Unless our total control design includes countervailing pressures such as these, the financial controls are likely to receive too much weight. Simply the availability of clear-cut standards and measurements favors their use, and it takes a financial executive with broad perspective to employ the tool with restraint.

EXPENSE CONTROL

Expense controls associated with finance deal with dollar outlays as shown in financial accounts. These outlays affect the flow of cash and the profits reported by the company, both subjects of direct concern to the financial executive.

Control standards can be set for every expense account maintained by a company. Although often based on past trends, the standards also are expressed as ratios to total sales or other volume measures. There are numerous standards of this sort, some comprehensive and some narrow in scope, penetrating every branch of a company. Moreover, normal accounting provides reports on actual expenses which can readily be matched against the standards. Here is a ready-made mechanism. And it is one that enables the financial executive—or other user—to become involved in all parts of an organization.

These convenient controls unfortunately have drawbacks which we must recognize if we are to employ them to maximum advantage.

1. They deal with inputs only. The outputs or benefits resulting from the outlays are not tied to the expense.
2. The expenses may be recorded in one accounting period but any benefits show up in later periods.
3. The control is post-action. The money is spent or other resource consumed before it is recorded in the expense account, and the report lags behind the recording. Orders, contracts, and other advance commitments, with rare exceptions, do not appear in the accounts.

4. When it is a financial executive who exercises control, operators often believe that he lacks an understanding of practical problems and is an illegitimate source of pressure.

Of the numerous kinds of controls, these financial expense controls probably create the most resentment. The negative reaction is due partly to difficulties we have just listed relating to legitimacy, time frame, post-action, and lack of tie-in to results. Also, expense reductions, of all external changes, often have the strongest impact, and the controls are blamed for the ensuing discomfort or inconvenience that result. Too often the resentment from those inevitable sources is heightened by *arbitrary adjustment* of expense standards, based on a desire for improved financial results but lacking appreciation for the operating sacrifices that will be necessary.

Financial expense controls are so convenient that they belong in every total control design. But we must minimize the negative response to them. Three approaches are suggested by the preceding discussion. (1) Treat financial standards and measurements as intermediate or second-level controls. Don't use them directly on operators or first-line supervisors. Instead, translate the desired dollar figures into operational terms—people, things, actions; assess the feasibility of these operational standards; and seek steering-controls that will lead to the results sought in the post-action controls. (2) Unless financial executives have the capacity and willingness to undertake the translation specified in (1), keep them in the background. (3) Watch the balance. Develop ways of monitoring the results of expenses along with pressures for expense reduction, and direct attention to long-run results of short-run changes. If an expense is important enough to control, it is important enough to control correctly.

FINANCIAL BUDGETS

Financial budgets are the best integrating device we have in our arsenal of control tools. The budgets are a forecast, or plan, of the figures in our profit and loss, balance sheet, and cash flow accounts at some future date. These projected figures become the control standards against which actual results are compared. The control standards are integrated or fitted together in the same way as accounting data are consolidated into monthly and annual financial reports.

Reasons for use. Three major advantages of financial budgets are (1) Use of the dollar as a common denominator for relating hours of Judy's work, TV commercials, gallons of gasoline, and innumerable other inputs and outputs. Being a common denominator, the dollar values can also serve in allocating resources. (2) Use of the existing accounting system with its wide coverage and established flow of data. (3) Direct concern with one of the key resources—capital—and one of the primary objectives—profit. No other control mechanism possesses these advantages.

Care is necessary, however, if we are to realize the full potential of financial budgets. The expense standards and norms for balance sheet items generated by budgets do not differ significantly from the standards already considered in this section. For operating control they need to be translated into actions and physical inputs and outputs which are meaningful to operators, and for balance they should be combined with other controls that focus more on results. It is the comprehensiveness of financial budgets that offers unique control possibilities.

Impact of preparation on acceptability as control standard. The way the budgets are prepared determine their usefulness as control standards. Three distinct approaches are available: estimates, financial allocations, and financial results of operating plans.

1. *Estimates.* When the chief purpose of financial budgets is to anticipate cash requirements, earnings, and similar financial items, carefully prepared budget estimates will suffice. Someone familiar with company operations and with past data can forecast what will probably happen. If necessary, two or three sets of estimates based on different assumptions can be prepared. By keeping these estimates in the form of profit and loss statements, balance sheets, and sources and applications of funds, the estimates can readily be related to past trends.

Useful as such figures are for financial planning, they have little value as control standards. If actual results deviate from the estimates, we don't know whether the estimates or the actual figures are askew. The people whose work is being measured feel no commitment to meeting the estimates and will feel annoyed if the burden of explaining deviations is placed upon them.

2. *Financial allocations.* Budgeting can be viewed as a process of dividing up available resources—as is done in a government, university, or community chest. Control is then focused on keeping actual expenditures for various purposes within their respective

allotments. This process is comparable to capital investment control, discussed on pages 86 to 88, and has similar drawbacks.

Although requests for allocations may be justified in terms of expected results, control activity is confined to limiting expenditures. Relating actual results to the outlays is not part of the process. And partly because accountability for efficient use is lacking, requests for allocations are likely to be large. This in turn invites playing games with arbitrary cutbacks and padded requests.

The breakdown of the budget deserves attention. Often the subdivisions are in terms of who gets the money—i.e., raw material supplier, salaried salesmen, electric utility, and the like—and are not related to the purpose to be achieved. So-called "line budgets," which may list each job on a separate line, are common in governments and universities. When such lines also serve as control limits, operators face great inflexibility; resistance and attempts to circumvent the budget are inevitable. Of course, making allotments in larger blocks, and preferably according to purpose rather than recipients, relieves this drawback, although it also reduces the regulation exercised by the budget administrator.[9]

A further drawback of financial allocation budgets is *when* control is exercised. The yes-no test may be too late. Often plans are all laid and commitments are made before the budget check on disbursements becomes effective. And in an on-going operation, we cannot shut down the elevators on the 24th of the month just because the budget allotment is exhausted.

3. *Financial results of operating plans.* A third approach to financial budgeting starts not with financial statements, but with action plans. The plans are concrete, such as selling 325 units to customers in Wyoming, phasing out the office in Houston, increasing direct-mail advertising to banks, hiring five more programmers, and so on. These operating plans are then translated into their accounting effects, preferably with subaccounts matched to organization units. This first accounting reflection of preliminary plans can be checked for internal compatibility and for acceptability. If revision is needed, instead of an arbitrary change in the financial figures, the operating plans are reexamined and only when they are modified is the budget revised.

Obviously, operating people must participate actively in this

---

[9] As pointed out on pages 134 to 135, federal government attempts to budget by "programs" via the Program, Planning and Budget technique have met only limited success. Line budgets continue to be used for repetitive activities which are difficult to separate into distinct programs.

planning cycle. Because the resulting budgets are more realistic, better understood, and partly homegrown, they will be accepted more readily (and perhaps with more commitment) as control standards. And when a comparison of actual figures with budgets indicates that corrective action is appropriate, a reversing of the translation from action plans to dollar figures should be clear.

For purposes of managerial control, then, budgets built up from operating plans are far more effective than estimates and allocations.

Need for additional controls. Even if they are tied closely to operating plans, two inherent drawbacks of financial budgets must be kept in mind. (1) Accounting records do not reflect a variety of intangible factors, such as morale, good-will, market position, product obsolescence, and the like. When to capitalize a disbursement is debatable. The fiscal year may not fit work cycles on, say, new products. And since financial budgets are tied to fiscal accounting, the budgets include these distortions.

(2) Financial budgets overemphasize short-run profits. We have mentioned in several connections that the explicit, numerical controls attract the most attention—be they salesmen's expense accounts or annual profit. Such obvious results are not always the only considerations in the long-run health of an enterprise, and certainly profit is an inadequate measure for a hospital or university.

As we develop financial budgets for control, then, we should also devise other standards and measurements that offset the inherent bias of the accounting data. Since we must not forego the benefits of budgets, the challenge is to find better measures of intangible and of probably long-run results.

Conclusions regarding financial resource control. Controls on the uses of capital are among the best developed and most powerful measuring and feedback devices we have. They should be a key element in every managerial control structure.

The troublesome issue is how to avoid stressing financial results too much. Because they are tied to financial reports for outsiders, and are therefore intentionally "conservative" and "objective," financial controls are sluggish in flagging opportunities or trouble in intangible areas and in signaling long-run impacts. This lopsided emphasis is especially critical if financial staff people enter deeply into operations. The short-run accounting numbers tend to get too much attention.

To maintain a reasonable balance, financial staff should be

kept in an advisory role to operating executives who have overall accountability, and additional controls should supplement the strictly financial measurements.

Conclusions: Resource Controls

A basic way to control any activity is through the resources needed. These inputs—the tools, fuel, services, supplies, manpower, capital—make results possible. By controlling them we exercise influence and power.

Consequently, every enterprise should impose various controls over the flow and use of its resources, as indicated in preceding pages. Although such controls must be fitted to each situation, several central themes emerge from the analysis in this chapter.

CONSERVATION VERSUS USE OF RESOURCES

As with national resources, each company has conservationists and activists. The conservationists seek to protect and regulate, whereas the activists, looking at the same resources, think of ways to use them, to promote change, and to fill new needs. This distinction between conservationists and activists is highly significant in designing a control structure.

1. The *overlap* of resource controls and other kinds of controls should be acknowledged. Materials control and project control, for instance, bear on the same set of activities. Likewise, financial budgets and control of a sales expansion program provide two perspectives on the same efforts. Moreover, manpower and facilities controls deal with inputs for the same operations as are covered by financial controls. This redundancy is inevitable; our task is to discover when it can be more help than hindrance.

2. Resource controllers will (and should) hold *different values and viewpoints* from those of doers. The resource people appropriately focus on protection, cautious use, and future capability. The doers want action and results. We should expect these values to clash since overlap exists in the objects being controlled. For example, the doer will want workers assigned where they will contribute the maximum to the results he seeks; the human resource controller will be more concerned with personnel development than short-run results.

3. Such competing controls should be created only when (a)

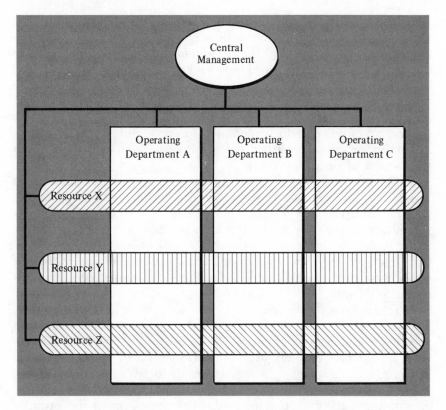

FIGURE 6-3. *Resource Controls Inevitably Overlap Program and Other Operating Controls.*

there is strong need to *assure adequate attention* to a particular resource, and/or (b) the activity being controlled is of such importance that a *double check* is warranted.

4. Insofar as we do set up competing controls, we should then also establish *mechanisms to resolve the conflicting pressures.* Morale suffers especially when competing demands arise from several different controllers.

ORGANIZATION FOR RESOURCE CONTROL

The negative reactions often provoked by resource controls can be relieved by wise organization. The role played by, say, a personnel director should differ for control of (1) supply, (2) alloca-

tion, and (3) use of manpower. Typically, supply problems are
separate enough from concerns of other executives to make inten-
sive attention to supply by the personnel director desirable. Allocation, by contrast, supports or holds back all sorts of action in various
departments. Here, a personnel director may advise, but his preoccupation with one kind of resource does not qualify him to decide
which control activities deserve priority. And with respect to use,
his control efforts should be confined to encouraging practices such
as counseling that strengthen the resource without detracting from
the pursuit of operating results.

Generalizing for all sorts of resources, controls related to
supply can be exercised by specialized divisions, but for allocation
and use line executives should dominate and resource specialists
should act as their service staff. Controls dealing with conservation
and efficient use of resources should be part of the package of controls employed by line executives. These executives are then in a
position to insist on balanced performance. They mediate the competing pressures.

Yes-no controls on allocations can also be made less irritating.
Basic allocations must be decided centrally to obtain coordination
and broad perspective. However, by stressing programs, large projects, and annual budgets, this central review can assign resources
in large blocks and for substantial periods in advance. Then within
these limits local managers can move without the necessity of returning for approval of each step. Under this arrangement the primary
role for the resource specialist is to assist local managers to achieve
agreed-upon results with the assigned men, equipment, money, and
the like. Such localized controls as may be necessary are steering
ones.

Clearly implied in these conclusions is that resource specialists
should not attempt to exercise control over the use of their respective inputs at local levels. For instance, financial experts should not
try to administer detailed expense controls. Rarely, too, does the
person who gives concentrated attention to one resource also understand the many local facts well enough to provide constructive
guidance. He may suggest a control device for the supervisor to
employ; and he may have a control (preferably a steering-control)
on aggregate results. If he attempts more, he is likely to create
resentment toward his entire function.

The resource controls that induce the strongest positive response focus on improving inputs and on technical assistance in
utilizing such inputs to achieve operating results.

FOR FURTHER READING

BOWER, J. L., *Managing the Resource Allocation Process*. Boston: Harvard Graduate School of Business Administration, 1970.
Four case studies of capital allocation in large corporations. The actual process was found to be much more diffused and complex than capital budgeting theory assumes.

COREY, E. R., and S. H. STAR, *Organization Strategy: A Marketing Approach*. Boston: Harvard Graduate School of Business Administration, 1971.
Summary chapters explore the conflicting interests between marketing program management and "resource" management.

HOFSTEDE, G. H., *The Game of Budget Control*. London: Tavistock Publications Limited, 1968.
Leading book that deals in practical terms with behavioral aspects of budgetary control.

ROWE, D. K., *Industrial Relations Management for Profit and Growth*. New York: American Management Associations, 1971, Chapter 24.
Guides for construction of a general personnel index.

VATTER, W. J., *Operating Budgets*. San Francisco: Wadsworth Publishing Company, 1969.
Clear, simple statement on budget preparation.

WELSCH, G. A., *Budgeting: Profit Planning and Control*, 3rd ed. Englewood Cliffs, N.J.: Prentice-Hall, Inc., 1971.
Comprehensive treatment of budgeting from accounting viewpoint.

YODER, D., *Personnel Management and Industrial Relations*, 6th ed. Englewood Cliffs, N.J.: Prentice-Hall, Inc., 1970, Chapter 25.
Reviews ways of auditing manpower management.

chapter 7

CONTROL OF
CREATIVE ACTIVITIES

A CREATIVE idea is both new and useful. Novelty alone is insufficient; the idea has to serve mankind by making life richer, easier, more understandable. Although we welcome creative ideas in all areas, some sections or departments of a company are explicitly charged with being creative. Research and development (R&D) and advertising are two leading examples.

Control of such creative work faces two difficulties which at least in degree distinguish it from control of other operations. (1) Creativity is surrounded by uncertainty. Neither the time nor quality of creative output can be predicted with assurance. And with output uncertain the inputs of manpower, money, and other resources necessary to achieve a stated goal are subject to wide fluctuations. Consequently, control standards are hard to nail down.

(2) Tight control stifles creativity. Some restraints are tolerable—creative ideas *do* emerge in totalitarian states—but research data indicate that unfettered combinations of thoughts in the specific area of investigation help. Thus, imaginative thinking cannot be programmed or produced on order. Also, many creative people derive great satisfaction from the work itself and from the status that comes with success. So any control efforts must be particularly concerned with maintaining enthusiasm, and with winning acceptance of objectives. Emotional response to control efforts is critical.

Nevertheless, creative activities are too important to be left uncontrolled. Just as ease of control should not dictate what to include in our control design, so difficulty of control is a poor excuse for indifference. Control of R&D raises one set of issues, advertising another. A review of these two areas, which are important in their own right, will also suggest ways to tackle creativity control in other areas.

R & D Control

The main approaches to control of R&D activity are (1) progress review, (2) personnel reassignment, and (3) control of resource inputs.

PROGRESS REVIEW

The uncertainty of outcome does not imply that R&D is unplanned. Someone selects the problems on which research workers will spend their time. A key element in good R&D management is clear identification of *research objectives* which support company strategy. Normally these objectives will be in terms of new products, product improvements, better processes, or other desired results. Even if the research is knowledge-centered, the objectives should be explicit.

With these objectives as screening devices, a series of projects is selected. Each project proposal will indicate the purpose; the general method to be followed; the approximate time when the work can be done; and the manpower, equipment, and money resources that probably will be needed. Everyone recognizes that both the results and the costs are uncertain, and the degree of uncertainty is taken into account in selecting the particular projects on which work can begin. Basically, this is a planning and allocation process.

To secure a degree of control, allocations for a large project are made only one step at a time, or are subject to cancellation at various check points. These *progress reviews* will be scheduled in advance at *milestones* or when the allocation is used up, or they may be injected when new data (external or internal) suggest a change in direction. The review is akin to a yes-no control, except that the criterion for decision is not sharp. Instead, progress to date plus any external change is the basis for a decision to proceed, alter the course, or stop.

Development work on atomic heart pacers, for example, was subjected to three thorough reevaluations in one company—when a physical model was completed, when the results of animal tests were known, and when tests with humans had run a full year. Of course, within the lab many interim checks were made. The major progress reviews, however, included appraisals by outside experts (ranging from atomic scientists to microbiologists to surgeons). Less dramatic are the prearranged progress reviews of an experimental rat elimination program in a western grain state. In this instance, the reviews take place on a seasonal basis.

A vital feature is *participation*. Since the work is highly technical and unprecedented, at least the project director and perhaps the entire team play an active part in preparing the proposal and in the progress reviews. Knowledgeable colleagues as well as supervisors also participate. The interpretation of findings, possible roadblocks, and new angles to try are so numerous and assessing the odds of success so involved that a pooling of ideas and judgment is essential.

This high degree of involvement increases the understanding and acceptance of the project goal. Although judgments regarding continuation are necessarily subjective, they probably will not appear to be arbitrary. And the emphasis in the review is on the future rather than on measuring up to past estimates. All these elements in the situation reduce emotional objection to this form of control by the persons doing the work, although at times disappointment over treatment of a favored project will arise.

To be sure, progress control of R&D projects lacks performance pars and objective measurement. In fact, most of the information about progress to date has to come from those performing the work, and they may bias their reports on the basis of how they perceive the reward system. But it does keep the work directed toward company objectives and encourages self-evaluation by committed operators.

PERSONNEL REASSIGNMENTS

Progress reviews focus sharply on the work itself. To evaluate people on the basis of the results of a single project is unfair because of the high chance element involved. Success may be merely luck, and failure to reach a goal may reflect skillful and diligent study of an untenable hope. Nevertheless, some people are more competent researchers than others, and over a span of time this competence will probably show in results.

So over a period of, say, two to four years, we can use a form of management by results. This is a post-action control. (1) The individual who turns up several good results and/or a real "winner" is rewarded by recognition, higher pay, and greater freedom in future work. (2) The persistent but unproductive researcher is encouraged to look for other work where his talents are more likely to prove beneficial. In some laboratories such a person is first transferred to other kinds of work within the organization; if the fresh start turns out no better than the first, the person is assisted in finding a job with another company.

All sorts of judgments are involved. Rarely is a project a conspicuous success or failure—like Salk's discovery of a vaccine for polio. A whole series of creative increments is more typical. When one person builds on the work of another, who gets credit? What of joint effort? How serious is a missed opportunity? So we have to deal with "batting averages."

The scoring gets fuzzy when "professional" ideas of successful work differ from company goals.[1] And if a man "plays a good game" —i.e., does his work in what appears to be a skillful manner—he may prolong his tenure. Nevertheless, over time the top quarter and the bottom quarter of performers can be distinguished. That is the time corrective action should be taken (even in the face of friendships which certainly will have developed).

Such a reward and penalty system becomes widely known. The luck aspects are recognized. However, if fairly administered the system usually is accepted. (It is rational, and to the scientist-engineer rationality has an appeal.) Being known and accepted, the control's chief impact is that a whole array of people anticipate its action and adjust their behavior accordingly.[2]

CONTROL OF RESOURCE INPUTS

Financial budgets, expense ratios, output per man-hour, and similar control standards focus primarily on *efficient use* of resources. The control impact is on careful husbanding of the resource and on complete utilization in the most productive manner.

Such emphasis on efficiency is not suited to creativity in R&D. Efficiency should still be treated as desirable because availability of resources does limit the creative activity. But efficiency control gives the wrong emphasis.

From a resource point of view, we aid creativity (1) by build-

---

[1] Professionalization tends to stress methods rather than results, and the methods often become ends in themselves. Nowhere is this more evident than in the academic profession.

[2] A few behavioral scientists contend that individuals cannot be motivated to increase their creativity; these psychologists believe that some individuals have the ability and others don't. Even if we accept this proposition, control via personnel management makes sense. The aim is then to identify creative people whose own standards are high and batting average is good and organize so as to utilize them fully.

ing a capability—a bank of talent and facilities suited to areas wherein we want new and useful ideas, and (2) by allocating that capability to projects which are attractive in terms of chances of success and payoff when we find a winner. Having "placed our bets" we are well advised to let the horses run. Feedback of expense and budget comparisons is unlikely to move us closer to our primary objectives.

Several factors restrain the role that resource control can play. Typically, capacity to do creative R&D can be changed only slowly. Two to four years—often more—may be required to train a young engineer or scientist in a specific area. Complex experimental and test equipment often takes up to three years to install and break in. The social structure of a laboratory takes time to develop. Consequently, expansion takes time after the funds are allocated. Contrariwise, because of the difficulty and expense required to develop the R&D capability, companies (and governments) are loath to make sharp reductions.[3]

Total outlays for R&D, then, have high inertia. They respond slowly to annual fiscal budgets and even more sluggishly to monthly budgets. Although the total outlays for manpower and facilities do get translated into dollars, and these dollars become part of the financial budgets, budgetary control is effective only in limiting total expenditures, perhaps by broad units of organization.

The more specific search for new and useful ideas is done by *projects*. These projects compete for allocation of the existing R&D capability. They usually have a dollar price tag, but the significant decision deals with manpower (or occasionally facilities)—how key individuals will spend their time. And these projects are best controlled by progress reviews, as already outlined. The original expense estimate (or manpower estimate) is unsatisfactory as a control standard because: (1) it is admittedly only a guess, and all sorts of reasons can be advanced by people doing the work why the guess turned out to be wrong; and (2) much more significant than arguing about past expenses is the decision about the extent and direction of future efforts.

The progress reviews, which help keep R&D effort directed toward findings especially valuable to the company and which reassess the odds of success, can improve creativity far more than

[3] The aircraft industry is an exception to these observations. In this industry highly specialized engineers are hired and discharged as the backlog of governmental contracts fluctuates. To an unusual degree, society and the engineers personally maintain the bank of talent. As the great reluctance to cut back NASA staff suggests, the social cost of such a practice is high and creative output suffers.

watching past expense ratios. As one research director says, "For creative work, we get good people and then try to excite them about projects important to us. After that all we can do is help and hope."

Several minor devices are commonly used to keep expenses from running wild. Logs of how time is actually spent; recording the use of materials; and procedures to secure approval for travel, overtime, and purchase of equipment are typical. Usually persons assigned to creative projects have considerable freedom on such matters. Nevertheless, the process of keeping track of expenses (in physical, not dollar, terms) and of justifying extras results in more careful consideration. And the records are known to be available for analysis if expenses do run beyond generally accepted norms.

Summarizing. The preceding suggestions for controlling creativity in R&D work appear to be flimsy. Creativity is an elusive aspect of performance; fortunately, other dimensions of project control are more predictable and controllable, as can be seen in

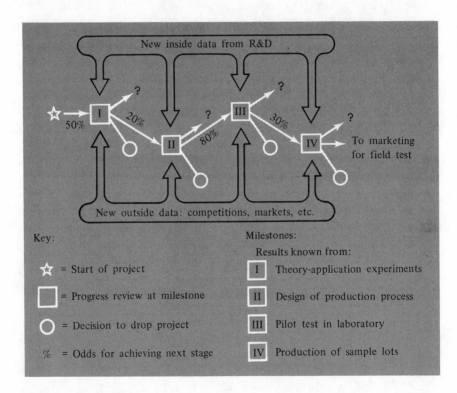

FIGURE 7-1. *Progress Review of R&D Project at Critical Milestones*

Chapter 5. Also, management can influence creativity by setting the proper stage through careful planning, and through alert mobilization and motivation of research personnel. The measurement and feedback controls on creativity, however, are inherently delicate and unreliable.

The chief methods of control operate at two distinct stages. In the short run, progress reviews at project milestones provide opportunity to reassess aims and prospects. This helps keep creative work related to agreed-upon objectives—it points up the "necessity" which we hope will mother an invention. And, with active participation of research personnel in the reviews, internalized commitment is increased. Thus, we have a form of steering-control.

In the longer run, evaluation of a succession of events enables us to separate the more creative people from the less creative ones. Then we reward and place our bets on those whose batting averages are high. Since the reasons underlying the less impressive performances are rarely clear, personnel reassignments which provide a fresh start for the poor performers are wise. But having made allowances for chance and for poor placements, the system should pay off in results. This is a clear example of post-action control. To maintain morale, these corrective actions (both rewards and transfers) must be based on a careful and systematic evaluation of long-run performance.

Control of Creativity in Advertising

Advertising has a recurring need for creative ideas. And, as in R&D departments, advertising agencies give systematic attention to incorporating creative inputs in their services.

Rarely do advertising agencies have separate units to generate creative solutions. Instead, they strive for creativity in promotion planning, in advertising themes, in copywriting, and in other phases of their work for clients. Each account has its own array of deadlines and expense ceilings, and these must be observed at the same time that creativity is sought. From a control viewpoint this means that several control mechanisms must act concurrently; measures of timeliness, of expense, and of creativity do compete for attention and management has a never-ending task of maintaining a reasonable balance between them.

This intertwining of creativeness with achieving output on schedule and within cost constraints is typical of many dynamic activities, from teaching to TV show production. So suggestions for controlling creativity in such situations have wide potential use. The brief description in the following paragraphs of steps that one advertising agency, Marsteller Inc., takes to control creativity does provide control concepts which have much broader applicability.

CONTROL POINTS IN THE WORK FLOW

At Marsteller, Inc., the advertising agency we are using as an example, creativity is explicitly and repeatedly stressed.[4] It is a prime consideration in recruiting and promotion, in financial rewards, and in procedures for handling work. This means that the controls related to creativity supplement a well recognized operating objective of the agency. Controls alone are not expected to produce results in this difficult and intangible area.

Nevertheless, the Marsteller system suggests several points in the cycle of generating client services wherein control of creativity can be exercised:

1. An *advertising program* is prepared for each client based upon an annual advertising plan. The annual plan sets the targets, and the program consists of preliminary proposals for themes, headlines, art, media, etc. This plan and program must be submitted to a Creative Review Board. This review provides an excellent opportunity for "steering-control." Plans are still flexible; time is available for change; individuals have not formed strong commitments based on long hours of work.

2. Following approval by a Creative Review Board, finished layouts and an entire campaign are prepared. When these proposals are ready for presentation to the client, they go through a *trial run* before a Plans Board. At this stage the thrust of the approach is rarely rejected, but changes may be requested in individual headlines or art. The control exercised here is primarily, though not entirely, concerned with meeting minimum standards and is of the yes-no type.

3. The status of total past relationships and future prospects for each client is explored at an annual Client Management Review. Naturally one part of this examination is an evaluation of creativity in the *work that has been done*.[5] This post-action control becomes

---

[4] This material is drawn from M. Anshen and D. A. Fuller, *Marsteller Inc.: A Case Series.* New York: Graduate School of Business, Columbia University, 1968.

[5] By this time some field results will be known. So many factors contribute to field results, however, that attributing success to the creativity of the advertising effort is still difficult, especially in industrial accounts, which are Marsteller's major clients.

valuable as it provides inputs (a) for the next annual program for
that client, and (b) for individual staff performance reviews.

4. All three of the above evaluations are client centered—they
are basically concerned with improving the quality of work done
for a specific company. Several hundred such reviews are held
every year. In this reviewing process some of the assessments—good
and bad—can be identified with the work of individuals. So these
individual contributions are summarized for the *performance* (and
salary) *review* for each person. Thus, the creativity controls are
indirectly tied to specific persons, even though the emphasis is on
generating outstanding work for clients.

USE OF SUBJECTIVE JUDGMENTS

The absence of objective standards is openly recognized. Examples of good and of weak creativity are cited, but with a warning
that each new situation calls for a different output. Whether a
specific proposal for, say, an advertising theme, a way of viewing
a client's products, or a script for a TV commercial is considered
to be "creative" admittedly depends on subjective judgment. Nevertheless, the *judgments are made* about every major program; they
are openly expressed and debated. From this continuing discussion
comes some consensus of the degree of creativeness demonstrated
in various pieces of work. However, the agency has not attempted
to translate this consensus into objective standards. The important
point is that the necessarily subjective nature of the measurements
of creativity does not deter their use. The control system is fitted
to a cardinal need, not to ready availability of numerical data.

Because the evaluations are subjective, *group judgment* is
widely employed. Each advertising program is examined by a Creative Review Board composed of immediate supervisors of the particular account plus representatives of at least two higher management
levels. Later a somewhat different group reviews specific proposals
that will be presented to the client. And annually three senior
executives participate in a Client Management Review. A major
reason for devoting so much executive time to these reviews is to
check subjective judgments about creativity. Also, the active participation by senior management helps to keep creativity an active
concern and to prevent the more easily measured factors such as
expenses and target dates from receiving a lion's share of attention.

ACCEPTABLE VERSUS OUTSTANDING OUTPUT

In creative work we cannot expect an inspired solution to every problem tackled. There are lesser degrees of novelty that may be acceptable. The way Marsteller deals with this range in quality is described by Anshen as follows:

> The management problem in controlling the quality of creative output can be best grasped if it is viewed as two related, but separable issues. The first was how to assure that all creative output met minimum quality standards—that no advertising or public relations work left a Marsteller office without satisfying threshold criteria. Clearly, this required (1) determining what such minimum criteria should be, and (2) establishing procedures for screening all work against the criteria. The second issue was how to advance the quality of creative output from minimally satisfactory to excellent. This was complicated by time or budget constraints that sometimes prohibited pre-testing of creative work, by inability to pre-test some creative work because of the character of the product or of its audience, by disagreements among skilled practitioners with respect to what was creatively excellent in specific advertising and public relations jobs, by the normal abrasions of human relationships within an organization, and by pressure to meet client or publication deadlines.[6]

Controls of the yes-no type can be set up for a minimum acceptable level because it is easier to get a consensus about quality at that level and because this kind of performance is always expected. The occasional outstanding output cannot be controlled in the same way since it occurs randomly and we are not prepared to reject work if it lacks this exceptional quality. In the short-run we have to rely on rewards for outstanding work rather than feedback controls.

REINFORCING CREATIVITY

What general concepts can we distill from this terse review of creativity controls at Marsteller Inc.? These observations are at least suggestive:

[6] Anshen and Fuller, *Marsteller Inc.: A Case Series*, pp. 194–195.

—Don't expect controls alone to assure creativity; instead, couple
 controls with careful recruiting, training, organizing, motivating,
 and other management actions.

—Do use subjective judgments about the success of creative efforts;
 admit that formal standards and objective measurement are not
 feasible and proceed vigorously with the best evaluations that
 can be made.

—To guard against individual bias and to improve the subjective
 judgments, use group evaluations of creative work.

—Distinguish between a minimum acceptable level of creativity expected of all work and highly creative ideas that blossom only
 occasionally. Screen rigorously for the former; encourage and
 reward the latter.

—Establish reviews of plans early in the work cycle when steering-
 controls can stimulate improved creativity.

—Set yes-no controls for threshold performance after operators have
 had opportunity for self-regulation but before external commitments are made.

—Confine post-action evaluations to assembling data for future use;
 little is gained from trying to invent a play for last Saturday's
 football game.

—Summarize a whole lot of (good and/or bad) creative inputs to
 individual performance reviews. Creativity is too ambiguous to
 pin a personal evaluation on one or two incidents. In the interim all individuals involved will be receiving feedback from
 the client oriented reviews.

GENERAL CONCLUSIONS

Control of creative activities—in sharp contrast to repetitive
work discussed in Chapter 4—runs up against uncertainty, subjective judgments, and prima donna responses. And the circumstances
under which we want creativity differ. The two areas probed in
this chapter, for instance, are unlike. Advertising campaigns must
meet deadlines and cost restraints whereas R&D projects can be
extended, expanded, or killed. Applied research and development
typically has a physical (and therefore measurable) objective
whereas advertising success is intangible.

Difficulty of control, however, is no excuse for abandoning it,
especially on a matter as vital to long-run success as creativity.
Several similarities appear in the R&D and advertising examples

just analyzed. In both, strong effort is made to maintain personal interest and commitment to explicit goals; the progress reviews of R&D projects and the frequent evaluations of advertising programs stress creativity. Participation is high in both evaluations. The short-run focus is on improving future results—i.e., steering-control. Also, the controls are similar in the minor attention given to methods. Significantly, the emphasis on personal acceptance of goals, participation, future guidance, and freedom from detailed check on methods are all features that we identified in Chapter 3 as contributing to favorable response to controls. For creativity, voluntary response is vital.

Protection from unreasonable competing pressures is at least recognized in both the R&D and the advertising examples. Other goals are clearly subordinated to creativity in the R&D control design. In the Marsteller scheme, time and cost are indeed measured and fed back, but a primary purpose of the various reviews is to prevent these measurable criteria from crowding out attention to creativity. Without some kind of protection, creativity is likely to become a casualty to controls on other more objective goals.

Both sets of suggestions provide for tough evaluations of individuals in the longer run. Success is rewarded; mediocre performers are removed. But fairness in such corrective action is stressed. Because of the inevitable chance in creating a winning solution, a whole series of observations are used in evaluating individuals. People are moved about, encouraged, given a second chance. Also, although judgment of success is often subjective, the measurement of each project is known and openly discussed. The aim is to have both successful and unsuccessful persons feel that managers lean backward to be fair, but expect and insist on high creativity over time.

FOR FURTHER READING

ANSHEN, M., and D. A. FULLER, *Marsteller Inc.: A Case Series*. New York: Graduate School of Business, Columbia University, 1968.
Descriptive case series on the strategy and operations of a successful advertising agency. Chapter 10 deals with "managing creative people."

PELZ, D. C., and F. M. ANDREWS, *Scientists in Organizations*. New York: John Wiley & Sons, 1966

Psychological research study of relationship between scientist's performance and the organization of his laboratory.

SEILER, R., *Improving the Effectiveness of Research and Development.* New York: McGraw-Hill Book Company, 1965.
Good book on management of R&D. Planning and budgeting receive primary attention.

chapter 8

CONTROL OF STRATEGY

THE master strategy of an enterprise is crucial to its continuous existence. The ability of the enterprise to carve out a distinctive niche in its environment is at stake. What services will be provided to which customers; how can these services be created at a competitive advantage; what major changes should be initiated and how fast; will the results of the activities thus prescribed be acceptable to persons whose cooperation must be obtained? These are the elements of company strategy. Unfortunately, the great importance of control in the strategy area is matched by its difficulty.

Control of strategy is difficult for three reasons: (1) The time span is long. Typically, several years are required to carry out a strategy, and we want to exercise control long before the final results are known. (2) Uncertainty is high. Competitors' actions, prices, governmental support, material supplies, and other aspects of the environment may veer from the expected course; also the outcome of one's own unprecedented efforts is hard to predict. This uncertainty makes control standards difficult to set. (3) The strategy itself may be altered midstream. Good strategy adapts to successes and failures, to new opportunities and new problems. So control must focus on a moving target.

Even recognizing these obstacles, the following five control measures are practical. The first two *signal* a need to revise our existing strategy, another two deal with the *success* in executing a selected strategy, and the fifth focuses on the *process* used in formulating a strategy.

Control Based on Updated Forecasts

Control of a spaceship heading for the moon illustrates the concept of steering-control—as we have stated in previous chapters.

111

Strategy control is even more tricky. The behavioral characteristics of the spaceship and the nature of its environment are stable and fairly well known. Not so for a social organization. The economic-political-social environment is dynamic, and the ability of a firm to follow a course is by no means certain.

For strategy control we can use the steering idea, but in a special way. After the early moves in a strategy have been made—which also means some time will have passed and the environment may have changed or reacted to our moves—the total situation is reassessed. We then prepare a new forecast of the likely outcome if we continue with original strategy. Estimates of likely output, costs, and timing are all updated, using newly acquired information. Perhaps ranges with probabilities attached to high and low estimates will be appropriate.[1] This updated forecast is compared with original targets, and—as with any steering-control—we decide whether to continue as planned or to modify the plan and/or targets.

Since a strategy is a complicated, multi-dimensional plan, preparing updated forecasts is an energy consuming and expensive task. But there is no satisfactory alternative. Partial forecasts of key items will, of course, be made more frequently; however, the strategy is a balanced and overall plan, and it should be assessed from time to time as a totality.

This evaluation of a strategy on the basis of an updated forecast is a recurring task. Whenever significant progress has been made and/or significant environmental changes might have occurred, the cycle is repeated. In the process, successive modifications may be made in the strategy so that original targets become quite outmoded.

When to fully update the forecasts and to review the strategy is hard to decide. One possibility is to tie the strategy review into annual revision of long-range plans.[2] This helps establish a planning routine which is understood by the many individuals who provide inputs. Unfortunately, the annual review date may fall in the middle of a critical phase or it may delay action that should have been

---

[1] A full analysis may also include risk exposure and practicality of contingency plans.

[2] Here "long-range plans" refers to three- to five-year financial budgets which are revised and extended annually. Strategy is more substantive and more selective in content; it includes projections over a longer span, and it gives more attention to ranges of possible outcomes and to contingency plans. Strategy and "long-range plans" can be developed together, but experience indicates that necessary attention to the many figures included in financial budgets tends to detract from imaginative thinking which is so important in strategy formulation.

taken two months earlier. An alternative timing is to use milestones —discussed in Chapter 5 in connection with project control—for example, the end of a pilot run, just before investing in facilities, or when volume reaches a level that suggests reorganization. Flexibility in timing the reviews is desirable if the critical junctures can be anticipated far enough in advance to prepare for the review meeting.

Monitoring the Environment

Uncertainty about changes in the environment is especially troublesome in strategy control. The strategy is based on key assumptions—planning premises—about wars, price changes, new technology, competition, legislation. and the like. If these premises are wrong, the strategy may have to be modified.

An essential step in any updating of our forecasts of the outcomes of our present strategy (for steering-control) is a *reexamination of the planning premises.* Notice that these premises are themselves forecasts of external conditions that will prevail perhaps three or more years hence.

The key planning premises of a firm selling machines and programs for "programmed teaching," for instance, were the following:

1. The education industry is so large and inefficient it must find ways to increase the output per teacher.
2. However, public schools will lag in this development because of their size, conservatism of school administrators, and the entrenched position of teachers' unions.
3. Instead, industrial training will pioneer, because industry is accustomed to think in terms of cost/benefit analysis and is more flexible in allocating capital for new methods.

On the basis of these premises the firm chose to concentrate on industrial training. A shift in any one of the premises, if recognized, would probably lead to a change in company strategy. Unfortunately, the firm is very vulnerable because its executives are completely absorbed with industrial training and to date have failed to establish a monitoring scheme to check the continuing validity of their strategy assumptions.

On the other hand, an engineering firm focusing on new processes to extract gasoline and fuel oil from Colorado oil shale, keeps a sharp eye on its external environment. Currently, the top execu-

tives predict that coal gasification will prove to be very costly, and
that Mid-east oil-rich countries will maintain their monopoly pricing on natural crude oil for at least ten years. These premises serve
as the basis for a strategy of concentrating on oil shale. However,
the risks are recognized. One engineer devotes full-time following
developments in coal gasification, and a political consultant reports
regularly on the Mid-east situation. Strategy is reconsidered whenever either of these monitors reports a significant change (and also
at milestones in their own work).

Some parts of every company's environment come under frequent review in normal operations. Union relations, for instance,
provide an ever-changing base for predicting future labor supply.
Although preoccupation with current issues may cloud the longer
view, the need to think about longer-run conditions will be recognized. Other assumptions often receive scant attention. New
technology and capacity being developed by companies not now
competitors, possible expropriation of raw material supplies, or a
shift in sales promotion media, for example, may be overlooked unless someone is specifically charged with watching for such changes.

So, key planning premises should be identified. These deal with
external conditions which (1) are subject to change and (2) would
upset the strategy if changed. Then explicit provision should be
made for regular monitoring of these areas. Revised forecasts should
be submitted for each major strategy review. In addition, if a major
deviation from the planning premise appears likely, consider scheduling a special strategy review.

Since managers cannot regulate these environmental conditions, control becomes a matter of spotting changes and adjusting
to them.

External Standards for Measuring Success

The two preceding approaches—updating forecasts and monitoring planning premises—provide the basis for steering-control of
strategy. How about post-action control i.e., evaluating the success
of a strategy?

Standards suitable for appraising the success of a strategy are
hard to find. Targets are in plentiful supply; each time the strategy
is revised a new set of targets emerges. But which target is the appropriate standard? Two or three revisions are often made before
results of the first actions can be known. The outcomes we observe

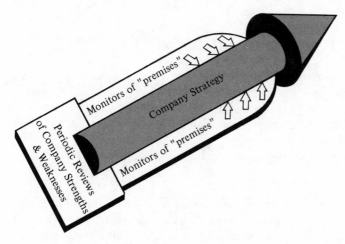

FIGURE 8-1. *Steering Controls that Help Guide Strategy.*

today flow from a succession of strategic plans; they reflect the wisdom of the planning, the skill of execution, and a host of environmental conditions.

A partial solution is to use the performance of similar enterprises as a standard. Since all the enterprises operated in the same environment, they presumably had similar opportunities and obstacles.

The simplest norm is a comparison of annual financial results, such as profit as a percentage of sales and profit as a percentage of total assets. Unfortunately, the simplicity is also a weakness. A limitation of these comparisons as a tool for managerial control is that the consolidated financial results are affected by so many influences that a tie-back to managerial action is extremely difficult.[3]

Comparisons of market penetration, production costs, new products introduced, and the like, are more revealing. They can be related to actions or inaction and they may provide some early warnings. Of course, the various companies will have pursued somewhat different strategies, but if our company failed to achieve relative success in its *high priority areas* we know that something was lacking.

---

[3] Other limitations arising from accounting conventions and from a narrow time-span have already been discussed in Chapter 6.

Control of the Process

Many enterprises are so unique that even crude comparisons with other firms are inappropriate. So, lacking a better standard, we turn to the process followed in establishing strategy. The presumption here is if the correct procedure is followed, the outcome should be good.

An evaluation of the strategy formulation process should include:

1. What *steps* are systematically taken? Careful examination of the outlook for each industry and identification of attractive niches; recognition of critical factors for success in each niche; matching company strengths and weaknesses against success factors; seeking synergistic benefits; determining resources needed to implement a selected strategy; developing a program of major moves; translating the program into targets for results—these are the steps in one approach.
2. Are qualified *people* doing the work? Here questions of breadth of interests, technical knowledge, imagination, attitude toward risk, and judgment are crucial.
3. Does the *reward system* encourage strategic action? Too often praise, bonuses, and promotions are based almost entirely on short-run profits. The executive who wisely devotes energy and resources for future benefits may be penalized instead of rewarded.
4. Is the *organization* designed to facilitate the necessary communication for strategy formulation and the effective execution of strategic decisions?

So much intuitive judgment is involved in designing good strategy that favorable answers to the above questions will not ensure excellent results. However, a strong set-up improves the odds for success.

Reaching Back

None of the strategy controls suggested above ties strategic decisions to individuals. Personal accountability is absent. This lack is serious because controls on short-run results are often strong, and individuals give much closer attention to areas that are more tightly controlled. A cavalier attitude toward strategy decisions can easily develop: "Make it look good. The plan will be changed three or four times before results show up, so why worry now? Besides, I'll be transferred long before the day of reckoning."

To overcome this tendency, a few companies "reach back" in time to tie specific decisions to specific individuals. Three steps are involved. (1) For each key decision write down who decided and who approved. (2) Wait until results are "known." (3) Then reward or punish. For example, in a large telephone company, when a serious blunder in inadequate expansion of facilities became evident, top officials examined the records to find out who had made the decision and under what circumstances. In this instance, potential need for expansion had been pointed out by some (not all) of the staff reports, but the chief operating executive chose a lower figure. Five years after the decision he was retired early, from a different position. Similarly, a large electronics firm transferred a man out of a key job for a mistake he made four years and two jobs earlier. And in each of these instances the reason for the personnel action was circulated through the executive grapevine.

One company goes a step further. *Both* the operating executives and the corporate staff men involved must concur in writing on major moves. This record is then available for audit when results are known.

Obviously, such reaching back cannot correct the mistake. The impact lies in the attitude generated in other executives; namely, "It could happen to me, so I'd better take seriously my decisions that have a long-run effect."

The reaching back technique has significant limitations. Substantial historical analysis is necessary to pick out critical decisions and to reconstruct the conditions under which they were made. Did the people involved act wisely with the priorities and information available to them at that time? Unfortunately, mistakes are easier to single out than wise decisions, so the technique tends to be negative and punitive. If pursued too vigorously, executives will become too conservative and spend valuable time concentrating mainly on building a safe record.

Occasionally, reaching back can help overcome a casual attitude about strategy decisions. To be fully effective, however, it should be coupled with high rewards for persistently searching out attractive opportunities and for sagacious risk-taking.

Conclusion

A recurring theme in our review of various types of control is: Don't let ease of control dictate what's watched; instead, select important factors and design controls around them. Control of company strategy fits this advice. Strategy control is neither simple nor exact, but it can have profound effects on long-run success.

Systematic provision for (1) monitoring key planning assumptions and (2) updating strategy forecasts encourages steering-control. Then (3) external comparisons with competitors on financial results and on critical success factors such as product leadership, costs, or market position, although crude standards, do indicate past effectiveness. And to relate gains or losses to specific individuals, (4) reaching back helps maintain a balance between short-range and strategic values. Finally, since each of these devices has serious limitations, control over (5) the process of establishing strategy is highly desirable.

These arrangements lack the precision and objectivity of, say, the repetitive action controls discussed in Chapter 4. Nevertheless, the payoff for even a small degree of improvement is so high that strategy control should be a part of control design of every well-managed enterprise.

FOR FURTHER READING

AGUILAR, F. J., *Scanning the Business Environment*. New York: The Macmillan Company, 1967.
Describes processes followed by large and small companies to obtain strategic information.

CANNON, J. T., *Business Strategy and Policy*. New York: Harcourt, Brace & World, 1968, pp. 84–98.
Presents suggestive outline for "auditing the competitive environment." The discussion, like others cited on this page, is directed toward planning rather than control; nevertheless, it suggests important elements to monitor.

KATZ, R. L., *Cases and Concepts in Corporate Strategy*. Englewood Cliffs, N.J.: Prentice-Hall, Inc., 1970.
Text sections present full framework for analyzing and forming company strategy.

NEWMAN, W. H., and J. P. LOGAN, *Strategy, Policy, and Central Management,* 6th ed. Cincinnati: South-Western Publishing Co., 1972.

Chapters 2, 3, and 4 discuss the approach to strategy formulation tersely outlined on page 110.

III

CONTROL SYSTEMS

chapter 9

BALANCING THE TOTAL
CONTROL STRUCTURE

THE design elements and behavioral responses involved in all control cycles were examined in Part I. We then analyzed in Part II controls for different situations and purposes. Now the merging of a whole array of such controls into a set of interrelated systems will be explored.

Need for Balanced View

NARROWLY CONCEIVED CONTROLS

Examples of specific controls that have caused more trouble than they corrected are all too frequent. One firm found itself in serious labor difficulty because a new inventory control plan, coupled with an existing cost control system, led to sporadic layoffs just prior to a union contract negotiation. The new inventory control by itself looked fine; it coordinated production with sales and lowered investment in finished goods. But in terms of the total operations, the effects clearly were bad.

A retail store, to cite another case, placed a strict yes-no control on hiring new personnel as a part of a general campaign to reduce costs. For most departments this control caused only minor inconvenience. At the same time, however, the maintenance department was responding to another control—budgets—by shifting from contracting-out to in-company performance of some repair work, and this involved adding men to the maintenance payroll. The maintenance manager was caught between two pressures. Since the control over hiring new personnel originated in the president's office, he took the easy way out. He continued to contract-out the repair work—even though this appeared to be more expensive—and met the budget squeeze by scrimping on other work.

In other instances, the *absence* of controls over, say, product development or cultivation of new customers may lead to difficulties if strong emphasis is placed on filling a tough sales quota.

The specific control in each of these examples had a worthy objective, but its effect was undesirable because it was not properly integrated into the total control structure of the company.

COMMON CAUSES OF IMBALANCE

Among the difficulties in achieving a balanced control structure, four types of problems arise over and over again.

1. Easily measurable factors tend to receive too much weight, and intangible factors too little. For instance, a company often watches carefully the expense per mile of its traveling salesmen but makes no check on whether the entire trip was desirable or undesirable. Control over research and development expenditures is in much the same predicament. In the personnel field, turnover figures can easily be watched, but the hiring of men with high potential often is subjected to little direct control.

2. Short-run results tend to be overemphasized compared with long-run results. In a typical control system, the annual profit-and-loss statement of a division receives more attention than building for long-run growth. Customer good-will, employee morale, product development, and even maintenance of equipment rarely "pay off" in accounting reports during the period when expense on such items is incurred. Long-run results are uncertain, and even if we withhold appraisal for a period of years, causal relationships are difficult to trace. Consequently, it is difficult to design controls that do not stress the short-run—or even the past.

3. The relative emphasis desired in a control structure may shift over time. For example, at one stage in the grov 'h of a company dependable quality and assured delivery may warrant higher priorities than costs; as competition increases, expense controls then need more emphasis; later, product development may become crucial to survival. At no time should any of these factors be completely neglected, but a shift in emphasis may be very desirable. In practice, the tendency is to leave existing controls intact and not adjust them to new objectives.

4. The high status and/or the energy of a staff man who is concerned with controlling only one aspect of operations may upset a desirable balance. One company, for example, promoted an

aggressive executive who had been on the board of directors ever since the firm was a small, single-product enterprise. His new assignment was staff work on manufacturing costs in all plants, a subject dear to his heart. Unfortunately, his corrective action was so vigorous whenever cost reports showed even small margins above his tough standards that supervisors resorted to many unwise practices to avoid a blow-up with this crusty character.

These four sources of misdirected emphasis, although by no means the only causes, do indicate why a manager should examine his control structure in terms of its overall balance and emphasis. Such a balance can be achieved in the following way:

1. Start with the desired impact on individuals.
2. Select key variables for integrating control action, but localize the administration of other controls.
3. Simplify the organization of control activities.

Start with Desired Impact on Individuals

ANALYZE ALL CONTROLS BEARING ON INDIVIDUAL IN HIS PARTICULAR JOB

Managerial controls, as we have stressed over and over, are effective only to the extent that they influence behavior of individuals, especially the behavior of operators and of supervisors who guide operations. So to get a handle on the diversity of controls in an enterprise, the best place to start is with those individuals whose behavior is of concern to us.

Think of the bank teller, the school dietician, or the branch manager as an example. What controls should there be over his work? To create a balanced set of controls impinging on such a person we start with (1) the key *results* we want him to achieve. What is his mission in the organization? Then (2) for *each* key result area we ask a series of questions: (a) Should steering-controls, yes-no controls, and/or post-action controls be used? (b) How measure the characteristics to be watched? (c) Who sets pars for each feature measured? (d) Who does the measuring? (e) Who evaluates the result? (f) To whom are reports of the measurements and evaluations sent? (g) Who takes corrective action? In addition, to complete the picture (3) we find out what other controls a person in the job is subject to and how these operate.

Table 9-1 indicates this analytical approach in matrix form.

TABLE 9-1. *Matrix for Analysis of Controls on an Individual*

Key results desired from job	Control points and type of control	Form of measure- ment	Sets pars	Mea- sures	Evalu- ates	Receives reports	Takes correc- tive action
					Who		
A	—	—	—	—	—	—	—
B	—	—	—	—	—	—	—
C	—	—	—	—	—	—	—
D	—	—	—	—	—	—	—
Other controls bearing on job							
X	—	—	—	—	—	—	—
Y	—	—	—	—	—	—	—
Z	—	—	—	—	—	—	—

The resulting picture provides a basis for assessing the balance of controls on that person in relation to his mission. The exercise of such controls—that is, the actual pars set, the corrective action taken, and the like—is a further issue. However, data suggested by the matrix are the main elements for design purposes.

MATCH CONTROLS WITH DESIRED RESULTS

The emphasis that any particular aspect of a job deserves should be clear from the mission assigned, which is derived through a chain of delegations and redelegations from company goals and strategy. If, in fact, this mission is not well understood, then a clarification of the job must be worked out before the control design can be settled. For short-term adjustments in strategy and tactics, the pars can be varied, as suggested in Chapter 2.

E. C. Schleh correctly stresses the prime significance of end *results*—the output of a person's efforts.[1] In Part II, however, we saw

[1] See E. C. Schleh, "Grabbing Profits by the Roots," *Management Review,* July 1972.

that some controls focus on inputs of *resources* (money, materials, manhours, machine usage, and the like) and some on *methods* which are expected to lead to results. These latter controls may be necessary and useful, but in our balancing of controls we should retain their subordinate role. Watching activities—say, calls made by a salesman—may be useful for steering purposes but its role as a means to an end should not be forgotten. Likewise, saving resources is an unworthy goal as such—instead, we want to use resources. In our controls the emphasis should be an effective use relative to results.[2]

The emphasis (or lack of it) on a particular phase of work can be generated partly by control design and partly by control implementation. In design, steering-control early in the action cycle, frequent yes-no checks throughout the cycle, and specific quantitative measurements increase attention. In implementation, participation in setting pars, prompt and consistent attention to deviations by managers, and strong reward/punishment tied to success raise attention.

LIMIT TOTAL CONTROL PRESSURE

Restraint is necessary. Increases in the total control pressure on an individual soon have diminishing impacts and can reach a level at which all controls are resisted. Also, emphasis is a relative matter; the response to controls in one area depends in part upon the push being exerted in other areas. Consequently, in balancing controls we should give just as much thought to reducing the stress on some aspects of the work as to increasing the stress on other results.

Control pressure on a selected activity can be reduced in three ways. (1) An acceptable level of performance can be set which is feasible in ordinary circumstances. As long as this "satisfactory" level is achieved, no further attention is expected. (2) The process by which such satisfactory results are achieved can be left to the operator. In other words, we remove control on "how." Often the added risks arising from removal of control over methods are small.

2 Missions expressed in terms of desired end results often omit explicit mention of cooperating with people in related jobs and fitting into the united effort. Clearly some controls may be needed to assure coordination among interdepartment activities; also, consistency of behavior in union relations or pricing may be sufficiently important to warrant separate controls. Additional controls such as these typically serve as constraints; if a satisfactory level is achieved, no further push is exerted. They do not enter into the balancing problem unless the total body of constraints becomes unbearable.

(3) Through training and endorsement of desired behavior, norms of good performance are accepted by people who do the work. The operators and managers develop a "professional" attitude toward the activity. Self-control takes over. When this occurs, external controls are no longer necessary.

One way to think about balancing controls is in terms of 100 per cent of a person's effort. If we want more attention to activities A and B, there will be less available for C and D. So instead of increasing the stress only on A and B while retaining existing controls on C and D, we recognize the need for a trade-off. Reducing the pressure on C and D will permit a transfer of effort to A and B. Of course, the switch is much more complicated than turning a valve. Nevertheless, by utilizing the ways of increasing or decreasing control pressure suggested in the preceding paragraphs, relative emphasis on various results can be shifted without continually building up control pressure.

Executive jobs, as well as operator jobs, should be analyzed in the individualized manner just described. The summary picture of all existing controls, the matching of controls with desired results, and the limiting of total control pressure apply to all sorts of positions. The scope of work covered and the corresponding measurements will, of course, vary. But the concept of controls believed to fit specific persons applies at all levels.

Select Comprehensive Systems Cautiously

Analyzing the need for controls job-by-job, as recommended in the preceding section, identifies many diverse regulating devices. Obviously some way to simplify and standardize this array of controls would be desirable.

The idea of a single, all inclusive appraisal scheme applicable to all departments is very appealing. Such a scheme would enable central managers to dip down into the organization and to compare the performance of various departments. Also, for those who regard control as a necessary evil, a single comprehensive system imposes the burden on all alike; no favored few escape the restraint of the pervasive system.

Unhappily, man has yet to devise such a system. The dimensions of good results are too varied and the need for feedback too specialized for any single measuring device to serve management adequately. However, the lack of a magic formula should not blind us to the advantages of more modest schemes which help pull together and simplify control efforts. Included within a total control design should be some mechanisms which (1) help coordinate interdependent activities; (2) assure central managers that adequate attention is being given to critically important aspects, such as product quality; (3) utilize potential economies in the measuring and reporting process; and/or (4) tie operations control to major objectives.

FINANCIAL BUDGETS AS INTEGRATING CONTROLS

The best control mechanism available to *business* managers that cuts across departmental lines and permeates all divisions is the financial budget. By using records and procedures already in existence for accounting purposes, the observation and reporting of actual performance requires very little additional cost. The relationship between even the most remote expense and the broad profit objective is clear and logical. The use of the dollar as a universal common denominator permits comparisons of departments and trade-offs in allocations. And combining budgetary control with the "power of the pursestring" gives clout to budgeting processes. For these reasons, financial budgets have a key role in integrating more specialized controls into an overall structure.

Financial budgets also have limitations. (1) The generality of the dollar as a measurement unit—which makes it so useful for the purposes listed above—also makes it a poor unit for coordination. For example, a dollar sales budget fails to give the production control section the specific information it needs to schedule output. Similarly, an adequate labor expense budget does not assure availability of people with particular skills.

(2) Financial accounting has serious limitations in measuring output. Primarily because of its intimate tie with conservative financial reporting which does not attach value to training and education, improved products, better health, better maintenance, customer good-will, and a whole host of other intangibles, we have to rely on other measurements and controls in these areas.

(3) The lopsided strength of financial controls creates a danger of imbalance in the control structure. Because quantified financial reports are widely distributed monthly, and because of the power of the pursestring, budget reports are influential. Unfortunately, they are best suited to *restraining* the use of financial resources and

to short-run results. Unless we develop countervailing controls, short-run financial results become, by default, the goals that receive the lion's share of response.

Clearly, then, our comprehensive controls should include financial budgets but not budgets alone.

COORDINATING INTERRELATED ACTIVITIES

The control design should aid coordination. A simple example is specifications for parts. If we want the parts made in our Belgian plant to fit a U.S.-made power microscope in an Italian hospital, the standards for quality control must be the same for all production shops. Also, if we expect to keep inventory of common parts or finished products balanced and not too large, control of the flow of purchased items, processing, physical distribution, and delivery commitments have to be regulated at some central point.

To meet such coordination needs, some pulling together of control effort is necessary. We must examine the particular ways that performance of one activity-center affects other activity-centers, and vice versa. Then, at points of interdependence some method of reconciling the control pressures at each activity-center has to be provided. Often the people in the centers themselves make mutual adjustments. But when communication among centers is difficult, when their normal interests diverge (as in production and marketing), or when close coordination of three or more centers is involved, explicit integration of controls is wise. Perhaps synchronized standards will be sufficient. If performance often deviates from standard, as in foreign deliveries, a centralized mechanism for feedback and corrective action has to be added.

For major projects a PERT system can be superimposed on the operating centers. The launching of a new product or construction of a hospital, for instance, calls for a complicated sequence of events. To keep these in order and on time, a PERT network and frequent evaluation of progress is helpful, as indicated in Chapter 5.

All such coordination arrangements rely upon existing controls at the operating level. That's where the action is. Without dependable local management, the overriding synchronizing efforts will be of little avail. To secure coordinated results we single out from a larger array of local controls particular standards and reports which are consolidated to secure integrated results.

ADEQUATE ATTENTION AND SAFETY

Decentralization does not mean abdication. When an executive delegates authority, he personally continues to be accountable for results. As in financial transactions, if A lends $1,000 to B and B lends the $1,000 to C, the fact that B has passed the money on to C does not remove the obligation B has to A. Consequently, even though an executive making a delegation has high confidence in his subordinates, he wants some control—at least post-action control— over the results achieved by the subordinates. (In fact, the existence of such controls is a positive encouragement to increasing the degree and scope of decentralization.)

The primary protection a senior executive who has decentralized can employ is a good set of localized controls. He should make sure that balanced controls—arising from an analysis such as outlined in the first part of this chapter—have been designed and then check to see that the controls are being actively utilized. Localized control, rather than remote control, is much more likely to prompt desired behavior; so the emphasis should be on shaping and encouraging the use of self-government.

A few aspects of delegated work may warrant independent surveillance—in addition to local oversight. (1) When a potential error can cause great harm to the entire operation, separate control is wise. Examples of such situations include quality in pharmaceutical production, language in an international treaty, and the like. (2) Similarly, potential theft of valuable marketable assets calls for a double check. (3) If for some reason the senior executive doubts that local supervisors will give adequate attention to a particular objective, he may resort to special control. Thus, when companies first adopt a new policy regarding "equal employment opportunity" for blacks, a separate topside check may be necessary to overcome social inertia.

The double checking for safety or adequate attention typically is achieved by a yes-no control at one or more key steps in an operating cycle. As we already have observed, this kind of control is restraining in character; it does help prevent mistakes but lacks a push for positive action. Unfortunately, steering-controls—which do encourage early initiative to reach an objective—are poorly suited for use as an independent check. Unless the senior executive who is worried about safety or adequate attention is prepared personally to follow up on warnings raised by steering-controls, he has

to rely on someone else to respond. And this reduces the reliability
of the safety device.

So, again, we come back to post-action review of the way subordinates control their own activity. *Restraint* via yes-no controls
is possible, but to promote *innovative* action the senior executive
should, first, devote time helping to design local controls that stress
the objectives he is concerned about, and, second, establish a good
auditing and reporting procedure which enables him to monitor the
operation of the subsystem (but not specific events).

ECONOMY IN MEASURING, EVALUATING, AND REPORTING

A third broad reason for consolidating control activity is to
ease and cut the cost of controlling.

Centralized data processing.   The use of computers is probably
the most fashionable topic related to the expense of controlling.
Computers do, of course, have amazing capacity to manipulate data
and turn out reports rapidly. Standardized reports showing quantitative comparisons for all sorts of classifications can be made available at several locations simultaneously. Typical topics covered are
inventory size, labor and material costs, sales orders and shipments,
turnover, and numerous other ratios. The per unit cost of such reports is low; the total cost of reports, however, usually increases
when computers are introduced because the number of reports rises
so rapidly.

Unfortunately, desired changes in behavior are not directly
proportional with the number of control reports generated. Having
control data—even non-quantitative data—that the persons whose
behavior we wish to influence find relevant and meaningful to them
is more important than masses of figures.

The more sophisticated data processing systems (1) provide
only key summaries and unusual deviations promptly, and (2)
maintain a large data bank which can be "questioned" for more
details or analyses when wanted. Under this arrangement the primary control activity remains close to the scene of action. Control
design is affected only to the extent that the terms of measurements
are adjusted to standardized data which can be easily fed into a
computer.

Ratios tied to physical volume.   To expedite the settings of
pars, and to help keep them "acceptable," ratios which relate inputs

and outputs to *physical* volume of activity are helpful. Hotel and motel chains, for instance, use guest-days as a volume indicator and then compute all sorts of costs or receipts per guest-day. Most ratios will be expressed in dollars per guest-day, but some items, like maids or bellhops, can be expressed in hours.

If such ratios become the normal pattern for management thinking, pressure exists to express most control standards in these terms. Thus, a large drug wholesaler expresses virtually all items in its detailed profit-and-loss statement in "per line" terms (a "line" is a single entry on a customer's bill). "Lines" are more closely related to most inputs than are dollars of sales. When such a *meaningful volume* index can be found, the task of setting pars is significantly simplified; the par is automatically adjusted for the actual volume of work passing through the "plant." [3]

Deadly parallel.   Another way to simplify controls—this time at the evaluation stage—is to rely on "deadly parallels." Two or preferably more operating units are deliberately organized along comparable lines; then the results of one unit are compared with achievements by parallel units. The assumption is that external opportunities and problems are about the same for all units, so "if the Dallas unit can do it, why can't you?"

This deadly parallel concept is a powerful control among the twelve Federal Reserve Banks, the Bell telephone companies, community school systems, and in many other situations. A competitive pride leads each manager to try to look good relative to his peers. In fact, there is danger that too much emphasis will be focused on these aspects of performance from which comparisons are drawn and that desirable adaptation to local needs will be sacrificed. Obviously, if the deadly parallel is to be used for control purposes, the criteria and methods of measuring must be put on a single comparable basis.

In all these ways to simplify and economize—central data processing, ratios tied to physical volume, and deadly parallels—some adjustment in strictly local control designs has to be made in favor of the broader system.

---

[3] A further refinement is a flexible budget based on the volume index. See Figure 4-1, page 57. Here the amount of change in each kind of input or output is set for a percent change in volume. A few expenses will be expected to change in direct proportion to volume, but the standard per unit will not rise or fall as fast as volume.

TYING CONTROLS TO OBJECTIVES

Failure to relate controls to basic objectives leads to considerable ineffectiveness in control effort. The failure occurs at both the top and the bottom of the executive pyramid. Senior executives too often give their primary attention to profit controls alone. Dollar profits and often R.O.I. (return on investment) are compared with monthly budget targets, and questions are raised whenever actual results slip below the standard. This prodding about short-term financial results reverberates beyond the executive suite.

Meanwhile, at operating levels where goods and services are created, the control emphasis too often focuses on *resource expenditures.* Outlays for materials, overtime, or advertising are watched closely and any deviation from planned expenditures becomes the center of attention. In addition, independent controls on, say, product quality, equipment maintenance, or compliance with wage-hour fair labor laws will be introduced—but these are often selected on a "squeaky wheel" basis to ward off trouble.

With such poorly conceived control design, topside pressures undermine intermediate and long-run adaptation to new opportunities, and the lack of attention to desired results at the operating level leads to perpetuation of the status quo. Achievement of objectives becomes a foreign concept.

A comparable disregard of objectives exists in most governmental and educational control systems. In a university, for instance, virtually no systematic effort is made to measure and control output, and controls at the operating levels are predominantly on resource expenditures. As in government, "line budgets"—which list in detail the salaries and items to be purchased—are the main instruments for recurring control.[4]

A major effort in the federal government has been made to cut through the traditional system. The Program-Planning-Budgeting (PPB) technique starts with the selection of program objectives, then spells out activities necessary to achieve the objectives, and moves on to budgeting the resources required for planned activities. In concept, resource budgets have to be justified in terms of contributions to program objectives. In practice, such ties to

---

[4] Of course, in both universities and governments elaborate procedures exist for screening persons to be appointed to permanent positions. Note that these appointment controls are also resource—not result—oriented.

programs are hard to maintain. Large, established departments perform a mass of continuing activities which are only loosely related to specific programs; and line budgets continue to be used for these departments. Also, the severe difficulties of measuring program results has delayed the imposition of controls designed to relate actual outputs with resource inputs.[5]

Fortunately, in business concerns it is possible to build closer ties between controls and objectives.

1. Controls at the top level start with a translation of broad objectives into shorter-range steps designed to reach the objectives. This is a programming task, already discussed for narrower objectives in Chapter 5. These intervening steps—such as R&D on a selected new product, executive development for foreign expansion, or market penetration in Alaska—become the goals for major departments. Then the process is repeated. Each major step is subdivided into smaller steps which become the goals for sections within departments. With such subobjectives defined, we design controls to help us move along the path.

   In practice, laying out these means-end chains is complex. Always several different subobjectives are being pursued at the same time, and we have to reconcile conflicting claims for scarce resources. Moreover, major steps often take several years to complete, and they detract from short-run financial profits. So measurements that enable us to keep long- and short-run goals in balance are needed. Nevertheless, building these bridges among broad objectives and activities in operating units is essential.

2. Controls at the operating level must include measurement of results which clearly and directly contribute to the strategic steps identified in (1) above. A whole array of steering and yes-no controls may assist in achieving results, but some post-action evaluation of accomplishments is also necessary. The standards used in this appraisal of results should be derived from the company strategy and its intervening steps which are translated into subobjectives.

This exercise of tying controls to objectives almost always flushes out weak links. Whether we start at the top and work down

[5] See A. Scheck, "A Death in the Bureaucracy: The Demise of Federal PPB," *Public Administration Review*, March 1973, pp. 146–156, for discussion of problem in implementing PPB.

or move up through the means-end chain, obsolete, irrelevant, or partial controls are likely to be discovered. Refinement of the control design will be necessary. Our localized arrangement for each job is likely to need some adaptation, in terminology and measuring units if not in substance, to fit into the integrated scheme.

Summarizing. For a variety of reasons we may decide to modify the cluster of controls that relate to each particular job. Coordination, economy, adequate attention, and tying controls to objectives all lead to some adjustments in order to fit the individualized controls into a broader pattern.

Nevertheless, we should not become so intrigued with a glamorous master scheme that we lose sight of the fact that controls become effective only through guiding the action of individuals working in their respective jobs. If we expect to have an appropriate balance and constructive personal response, the controls should be tailored to each situation. Similar measuring devices will be used for similar tasks, of course, and other features of an overall system may be appropriate, but we will secure greater effectiveness by thinking of these common elements as aids to individualized controls rather than vice versa.

Simplify Organization of Control Activities

The effectiveness of any control system depends, in part, on who does the controlling. The more complex the organization, the more likely the controls will be distorted. Resistance and even sabotage is invited by multiple sources of control pressures. Three ways to help overcome such negative responses are to (1) emphasize the service role of resource units, (2) place control activities close to operations, and (3) have generalists balance controls.

EMPHASIZE SERVICE ROLE OF RESOURCE UNITS

The major source of friction in controlling revolves around the roles to be played by "resource" divisions; i.e., units charged with providing inputs of money, manpower, or services. We knowingly create this friction. The very reason for having personnel, engineering, finance, legal, product quality, maintenance, and other resource or service divisions is to be sure that adequate attention is paid to developing and protecting these necessary elements. But such divisions can be over-zealous. Our continuing task is to keep the specialized interests that we have devised in balance.[6]

Differences in values and in information are common. Each resource group naturally attaches high importance to nourishing and protecting its particular asset—as pointed out in Chapter 6. Operating people, on the other hand, are results oriented; they strive for ends and tend to exploit the means. Moreover, each person typically is better informed about his own tasks, and is more keenly aware of action needed to fulfill his own mission.

To help keep these differences in viewpoint within constructive bounds, our organization of control activities should include the following features.

1. *Select targets by mutual agreement.* Both the resource unit and the operating people should concur on factors to be watched and on results (or pars) expected. These should be predominantly results to be achieved in operating departments, such as labor cost per unit output, new employees trained, and the like. If acceptance is genuine, the control process is likely to work, whereas superimposed standards (except for trivial matters) will be resisted and perhaps sabotaged.

2. With targets accepted, the control activities of the resource and service units should *emphasize assistance to the operating departments.* The staff units can help monitor and interpret steering-control. They can analyze post-action results for the purpose of learning. The relationship becomes one like a lawyer serving his client rather than being a prosecuting attorney.

3. If yes-no controls are inserted in the operating flow and administered by an external division (e.g., checks on pay increases, quality, press releases, etc.), the *grounds for rejection should be explicitly and narrowly defined.* When a rejection is clearly necessary to attain an agreed-upon goal, the independent assessment may create disappointment but not rancor. A rejection based on a newly contrived standard, on the other hand, is sure to be resented.

4. In the routing of control reports, *always inform the person(s) who directly regulates the performance first.* He can often take cor-

---

[6] These resource and service divisions act in a *"staff"* relationship to *line* executives. However, because the meaning of the terms "staff" and "line" varies so widely in current usage, we have shifted to other language. Hopefully the more descriptive words—operating for line, and resource and service for staff—avoid the customary ambiguity. The terms are not fully synonymous but in most of the discussion in this book they can be read as if they were.

rective action immediately, or if necessary he can call on the appropriate resource unit for help. Psychologically, if he initiates the contact, or just feels that he probably should have done so, the chances of a positive response to the control are improved.

5. Although the resource unit is clearly cast in a service role, we can require each unit to submit an *annual report on the opportunities for improved use* by operating people of the resource and service he provides. The resource man should know how well his service is being utilized, and operating people should know that he is obligated to report "foot dragging."

These guides for resource unit/operating unit relationships do confine the power of resource units. However, the net contributions of such units will improve because resource control will be better integrated with purposeful action of the enterprise.

PLACE CONTROL ACTIVITIES CLOSE TO OPERATIONS

The chances of prompt and constructive response to a control are raised by placing the persons performing control activities close to operations. Typically, they should be close physically and organizationally. For example, the open-hearth furnace operations of a steel mill were significantly improved by moving metallurgists, who ran tests on the operations, from the laboratory to the furnace floor and by having them report to the shop foreman; both quality and through-put rose.

Reasons for close proximity. Benefits of having controllers close to those controlled include (1) faster feedback of control data, (2) evaluation based on local facts, (3) broader array of opportunity for specialists to participate in steering-actions and to be constructive in other recommendations.

Effect on expense and independence of control. The limitations on close proximity are partly real and partly contrived. Expense clearly prevents a full complement of controls in each operating section. So we resort to centralized data processing, part time on combined job assignments, less division-of-labor within the control process, and similar adjustments. But the aim is to physically localize control activity unless the expense of doing so is very much higher than centralized activities. A less valid objection to close proximity is that the controllers will be seduced by their under-

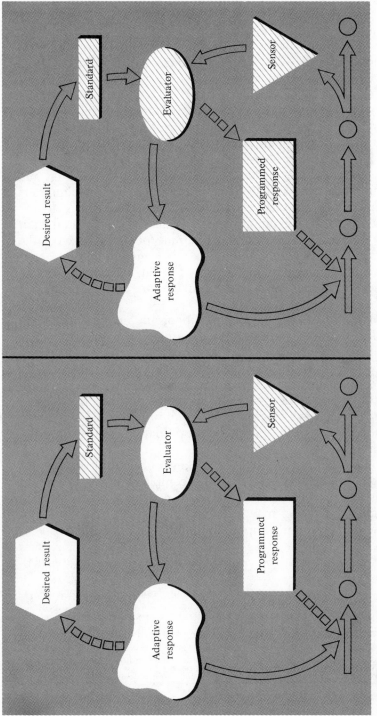

FIGURE 9-1. *Alternate roles a resource or service division can be assigned in a control cycle. In these two examples, the resource or service division performs the role in the striped spaces.*

139

standing and sympathy with local needs and will fail to enforce standards. Our argument thoughout this book is that the behavioral changes controls are really trying to induce will come more from working with operators to achieve standards than from strict external enforcement.

Persistence is necessary to retain a balance between the competence of specialists without large centralized control divisions. Specialists are usually more comfortable living together, speaking the same language, and cherishing the same values. But this kind of reinforcement for the specialist can be supplied by meetings and other communications, and by having a supporter at headquarters when promotions and salaries are being discussed. We can and should have functional relationships between engineers throughout the company, between accountants at various locations, between the lawyers, and between other specialists who perform control activities. However, to group all specialists of a particular training into a single division encourages narrow thinking and autocratic control.

Distinct role of auditing. One important exception is vital. The validity of control reports must be unquestionable. Managers throughout the organization must have confidence that the control information they receive has not been twisted or altered to cover up unwelcome news. So we maintain an *audit* function to make sure that the reported information is all that it purports to be.[7] And to attest to the integrity of data, the auditors should be separate and independent from the persons who prepare and distribute the control reports.

Internal auditors should verify all kinds of control reports, not just financial reports. Sales orders or compliance with pollution regulations may be just as crucial to good control as payrolls. Such an extension of auditing may be resented. Some people feel that any review of their work implies a doubt about their honesty or competence. So the essential need for auditing all sorts of reports on a regular basis should be carefully explained. Once established,

---

[7] Not all control reports are accurate. As pointed out in Chapter 2, promptness is often as valuable as a high degree of accuracy. We presume that executives will appreciate the quality of the data they are receiving. The auditors' primary role is to check to see that such reports have been prepared in the accepted manner and are consistent with previous reports. Typically, auditors make a post-action check to catch careless errors or manipulation; the expectation of an audit helps prevent bad practice. Auditors also make suggestions on how changes in procedures can improve controls.

auditing procedures should become a normal, impersonal verification.[8]

HAVE GENERALISTS BALANCE CONTROLS

Assigning primarily service roles to resource groups and placing representatives of such groups within operating departments, as recommended above, will simplify the control structure. These moves help integrate controls with primary operations; they put most of the feedback close to basic action. But the diversity of pressures remains. Each conscientious controller—be he concerned with money, manpower, materials, law, engineering, environment, safety, product quality, or other characteristics deserving attention —naturally stresses his particular mission.

Someone needs to assess the relative impact of the various controls and decide whether consolidated effort matches the priorities dictated by company strategy. This task belongs to the chief executive. He interprets the strategy, he allocates the available resources, he approves operating plans, he provides motivation and leadership—and the control system should reinforce his action in all these phases. Possibly he will have a general staff person (a "chief of staff") assist him in overall management, or an executive vice president who directs internal operations; and he certainly will have executives who manage large segments of company operations. These subordinates typically will assist the chief executive in balancing the controls. The essential requisites for the task are a

---

[8] The term "auditing" has several different meanings. In this book it simply means *verification* of management reports. The Institute of Internal Auditors, however, lists a much wider range of activities that may be performed by a person called "internal auditor." See V. Z. Brink, J. A. Cashin, and H. Witt, *Modern Internal Auditing*. New York: Ronald Press, 1973. Brink argues persuasively that an internal auditor should (1) check compliance with procedures, policies, and plans; (2) verify management information; (3) protect assets—especially from fraud and dishonesty; (4) appraise the adequacy of controls; and (5) contribute to the appraisal of performance and to recommendations for operating improvements. Although in large dispersed companies such a broad scope for the internal auditor may be desirable, we have stressed in earlier chapters that primary responsibility for appraisals and compliances should rest with line executives and specialized staff. Information and insights from internal auditors will, of course, be welcome. But attention to such matters by auditors should not detract from their unique and vital functions of verification and protection from fraud. As with a public (external) auditor, the unquestionable capacity for independent verification should never be sullied by too deep involvement in other advisory activities.

grasp of the total mission and activities and an objectivity about the relative significance of the various features of the total effort.

These generalists have three vital functions in balancing the controls.

1. Checking the adequacy of the various control mechanisms to assure attention to all major objectives and subobjectives. (This book provides an approach for such an appraisal.)
2. Varying the weights to be attached to various objectives and consequently the pars to be sought by the corresponding controls. In many areas a "satisfactory" performance will be enough; in other areas exceeding the targets will be the goal. These weights will change from time to time depending on external opportunities and pressures and on the level of recent accomplishments.
3. Eliminating controls no longer needed. Controls tend to linger on—even after work habits are sufficiently strong to assure adequate performance without formal controls, or the emergency prompting the controls (e.g., a price freeze or power shortage) has passed, or a new control provides an adequate check.

Only persons with a comprehensive view of the management system should take these actions.

Fitting various phases of diverse controls into a comprehensive control structure has been the focus of this chapter. Starting with the controls relating to separate jobs, we next considered reasons for combining some of these activities into various sub-systems. Then guides for effective organization of control activities were reviewed.

Throughout this chapter the attention has been on controlling. Only occasionally have we dealt with the interaction between control and the phases of management—planning, organizing, and leading—yet we know that design in any one phase affects the others. Some of these interactions between control and other management phases are examined in the following chapter.

FOR FURTHER READING

BACON, J., *Managing the Budget Function*. New York: National Industrial Conference Board, 1970.
    Clear summary of the use of budgets for coordination and control, based on survey of company practice.

DEARDEN, J., "MIS Is a Mirage," *Harvard Business Review*, January 1972. Summary of the obstacles to a single integrated management information system. Similar difficulties for a single overall control system can be deduced.

HOLDEN, P. E., C. A. PEDERSON, and G. E. GERMANE, *Top-Management.* New York: McGraw-Hill Book Company, 1968, Chapters 2, 5, and 6. Central control in large enterprises; results are compared with those of a similar study twenty-five years earlier.

PRINCE, T. R., *Information Systems for Management Planning and Control*, rev. ed. Homewood, Illinois: Richard D. Irwin, Inc., 1970. Focuses on designing the internal information flow for planning and control purposes.

chapter 10

INTEGRATING CONTROLS WITH TOTAL MANAGEMENT DESIGN

THE intimate relationship of controls to other management processes has been indicated throughout our discussion. Objectives and other goals, for instance, underlie the selection of control standards; programs find their financial expression in budgets; decentralization and participation have a marked influence on the acceptance of controls; and so on.

Because the design of a control structure is so often glossed over, we have made it the center of our attention. However, many control designs can be made more effective by modifying the company's planning, organizing, or leadership style. Conversely, sometimes the control design should be adjusted so as to aid the planning, organizing, or leading. Obviously such trade-offs should be considered, especially if synergistic effects in the total management design are possible.

In this chapter, we have singled out several ways controls can be fitted together with other management processes. These opportunities do not begin to cover all of the interrelations. Rather, they are issues that arise time and again in actual practice, and resolving them wisely can be a great aid to effective management. These issues are:

1. Clarifying planned results to aid control.
2. Relating control to *new* planning.
3. Decentralizing without loss of control.
4. Harmonizing departmentation with controls.
5. Creating profit centers.
6. Adjusting controls to matrix organization.
7. Enhancing controls by leadership action.

Clarifying Planned Results to Aid Control

The basic concept that planned results become the goals of control is simple enough. However, to achieve this neat relationship requires continuing managerial attention. Deciding to increase the ratio of executives who are women or to obtain more stainless steel from European sources, for instance, merely states an end result. Since control becomes effective only through modifying the behavior of persons, these new goals have to be translated into more individualized standards.

Planning must be pushed from broad objectives to successively narrower and more specific tasks until each necessary move or component is assigned to a particular person. Then controls at the subsidiary level will contribute to the final result. If the complete plans for the construction of a new building, to cite a simple instance, are properly integrated, and then controls set up over the work of the foundation subcontractor, the structural steel subcontractor, the electrical subcontractor, and everyone else who has a

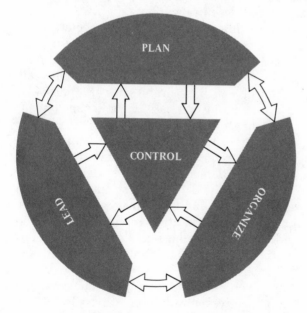

FIGURE 10-1. *Interaction of Different Phases of Managing. By integrating the subsystems, the effectiveness of the combined effort is greatly increased.*

particular part to contribute to the total structure, the final result should be a building as conceived by the architect.

Unfortunately, this sort of matching of the control structure with the company objectives is hard to achieve. Often the planning is not extended to the point where we can safely rely on individual discretion to complete the task. And if a new objective is unusual, our normal measuring devices may not reflect the distinctive features needed. So, two questions should be asked: (1) Who must achieve what results if the new objective is to be obtained? (2) How will we know that the necessary contributions to the final result are being made? *Planning is incomplete until concrete steps have been identified and provision made to control this implementation.*[1]

The elaboration of a plan down to results necessary from an array of individual operators or units need not be prepared by a single central planning body. A large block of the total work may be delegated to, say, the purchasing division, and the elaboration of plans for that block developed within the division. Such decentralization, however, does not reduce the need for full planning and subsequent control; only the location of who does the planning and control is changed by the decentralization.

Such elaboration of planning down to results expected of individuals and then creating controls to check on these results—as just recommended—is hard to keep flexible. The controls, once established, tend not to be readjusted as objectives are changed. Take even the simple matter of a cutback in the sales of a particular line of products due to a change in competition or technology. It is entirely possible that although the overall income and expense objectives will be adjusted to the new conditions, the control standards for engineering and other service departments will remain unchanged. Or suppose the president of a company decides to increase the number of broadly trained young men and women in the organization as a reservoir for filling top management positions. If the job specifications that control the representatives who actually hire college students are not adjusted, the specific actions at the various recruiting centers will not be attuned to the new objective.

The several steps involved in translating the new company objective into revised subobjectives for different divisions of the company, securing understanding and acceptance of these objec-

---

[1] Difficulties of relating control standards to objectives in government and education are briefly discussed on pages 134 and 135.

tives, adjusting the control standards and measurements accordingly, and using the control mechanism so that it actually influences behavior all take time and effort. In extreme cases, the inertia is so great that inconsistency between company objectives and the controls that are really ineffective continues indefinitely.

So the need to link planned results to control standards arises over and over. Whenever new plans are laid we should ask, "What corresponding adjustments in controls must be made?"

Relating Control to NEW Planning

Three types of control were identified in Chapter 1: steering-controls, yes-no controls, and post-action controls. Normally, steering-controls and yes-no controls are used to adjust activities so that a predetermined objective will be achieved. A thermostat turns heat on or off, bacteria tests in a milk plant flash warnings to the processing units and to the whole collection system, and so forth. Corrective action may involve detailed planning, renewed motivation, and other managerial acts. In this sense, control may prompt action in any of the other managerial processes. Nevertheless, objectives usually remain the same, and the adjustments are like those of a ship's pilot who modifies his course with the winds and tide to reach home port. Only if there is a terrible storm or breakdown is the pilot likely to change (replan) his destination.

In contrast, post-action controls almost always lead to planning. For example, if a sales campaign is only partially successful, both objectives and methods of the next campaign are likely to be modified; similarly, executive development activities planned for next year will be strongly influenced by an appraisal of results achieved this year. In situations such as these, control reports serve as a basis for an entirely new cycle of managerial activity—planning, perhaps organizing, leading, and controlling the new activities.

But this concept—that post-action controls rather than steering-controls are of principal use in planning—needs one important qualification. Often we must lay plans for new activities before a present cycle is completed. University budgets, for example, are often prepared in preliminary form in December and January for the following school year. This means that the results of the fall-semester activities are not yet known, and the spring semester has not even begun, when the first steps of planning for new courses and size of classes have to be taken. Automobile companies have an even greater lead time in planning for their new models; commitments on design and tooling are often made with little or no measure of the popularity of the current year's model.

When new plans must be made before the results of the old ones are known, the results must be predicted. Control information of the steering type is naturally used in making these predictions. In some situations, then, we use information on how we are doing both as a guide to current operations and as part of the data on which the outcome of present and new plans are predicted.[2]

Care is necessary when the same data are used for both planning and controlling. An over-zealous vice president of a furniture manufacturing company, for example, used some expense ratios projected for a new method of operation as a standard in appraising current activities; this discouraged and disgusted the men in the plant because they felt the new standards did not apply to their current operations. On the other hand, we may sacrifice accuracy for promptness in compiling control data, thereby limiting the value of our figures for planning. The purpose of a measurement should and does affect what is observed.

Decentralizing Without Loss of Control

Each time a manager delegates work (operating or managing) to a subordinate, he creates the problem of knowing whether the work is performed satisfactorily, and so delegating inevitably raises the question of control. Often the degree of decentralization a manager will adopt is tied to how far he can do so "without losing control."

Modify control as decentralization increases. A manager need not lose control when he delegates a large measure of planning, but he should be prepared to change his controls. This alteration is illustrated in Table 10-1. First, the types of control standard that are appropriate change. When decisions are centralized, the manager himself will establish rather detailed standards for the method and output of each phase of the work. But as he delegates increasing amounts of authority to plan and decide, the manager should shift

[2] The service of control to planning, however, should not be exaggerated. If we want planning to be dynamic, we must consider new ways of performing work. Operating conditions change, and future opportunities may improve or diminish; consequently, more complete, or different, information is often needed for planning than control activities provide.

TABLE 10-1 *Effect of Decentralization on Control*

Nature of Control

Degree of Decentralization	*Type of Standard*	*Frequency of Measurement*
Centralization of all but routine decisions.	Detailed specifications on how work is to be done, and on output of each worker.	Daily for output; hourly to continuous for methods and for quality.
Action within policies, programs, standard methods; use of "exception principle."	Output at each stage of operations, expense ratios, efficiency rates, turnover, and the like.	Weekly to daily for output; monthly for ratios and for other operating data.
Profit decentralization.	Overall results, and a few key danger signals.	Monthly for main results and for signals; quarterly or annually for other results.

his attention away from operating details to the results that are achieved.

The frequency of appraisals also changes. Because the manager is no longer trying to keep an eye on detailed activities, most, if not all, daily reports can be dropped. As his attention shifts, with increasing decentralization, more and more toward overall results, the span of time covered by reports can typically be lengthened. For a division that operates on a profit-decentralization basis, monthly profit-and-loss statements and balance sheets come as frequently as most top managers want reports. Other factors, such as market position or product development, may be reported only quarterly.

Retain safeguards.   The shift from frequent detailed control reports to periodic general-appraisal reports does not preclude the use of a few controls of the danger-signal type. A common practice is to expect a subordinate to keep his boss informed of impending difficulties rather than bother the manager with control data when conditions are satisfactory. A manager may ask to be notified when deviations from standard exceed a certain norm, thus applying the "exception principle" to control.

Moreover, yes-no controls can be used for certain major moves, such as large capital expenditures or the appointment of key executives. Here again, the number of proposed actions that require confirmation will decrease as the degree of decentralization increases.

Still another kind of safeguard is to insist that lower levels of management use specific control devices even though an upper executive himself neither sets the standards nor receives reports on performance. A vice president in charge of production, for example, may be vitally concerned that a reliable quality-inspection plan is in use, but he may take no personal part in its operation. He expects sufficient control data to be handy if the need for determining the cause of any particular problem arises.

As more authority is delegated, self-control by the subordinate becomes crucial. Such self-control is partly a matter of attitude and habit. In a situation wherein centralized control has been the traditional practice, operating personnel naturally rely on senior executives or their staff to catch errors and initiate corrective action. If authority is then passed down to them, they need to formulate a new attitude. It may also be necessary to redirect the flow of information so that these people down the line have what they need in order to do their own controlling.

With heavier reliance on self-control by subordinates, the manager should act more as a coach than the one who decides on corrective action. Ideally, the initiative for corrective action comes from the subordinate. To foster a relationship in which a subordinate is not reluctant to seek advice on tough problems, a manager should (1) avoid giving the impression that he feels an admission of difficulties is a sign of weakness, and (2) be careful not to make unilateral decisions that, in effect, take authority back from the subordinate.

Set the stage. The kinds of control and their associated relationships we are discussing grow only in a favorable climate. To create such an environment, we must think out a clear set of objectives for a task that is being delegated and develop ways of measuring the achievement of these objectives—as recommended above. Also necessary is a clear understanding about which policies, organization, management methods, and other company rules *must* be followed and which may be regarded only as recommended practice. Moreover, those actions that require prior approval by a boss need to be so labeled.

In addition to the substantial amount of planning and organization clarification just outlined, high decentralization requires the right people. Subordinates able to perform the delegated duties must be selected, trained, and properly motivated. An executive himself must be able and willing to adjust his behavior, and the two particular personalities involved in each delegation must trust each other. Remove or significantly diminish any one of these aspects of an operating situation and there will be a corresponding reduction in the degree of decentralization that is possible without loss of control.

Harmonizing Departmentation with Controls

The ease of control is significantly affected by the way the work of a company is grouped into departments and divisions. Important here are the concepts of clean-breaks, deadly parallel, and direct interaction.

The simplest way departmentation can aid control is by separating departments or sections where a clean break in work occurs. Thus, a farmers' buying co-op will separate bulk fertilizer, seeds, fuel oil and gasoline, and garden supplies; each requires distinct storage and delivery equipment. Likewise, a well-run ski resort will have separate divisions for its ski run, its ski shop, and its housing and food; to separate control of its restaurant from its bar, however, becomes more difficult because the service is so interrelated. Control is easier when either the physical separation of operations or distinct stages of work make it simple for everyone to understand the organization structure.

A second suggestion is to set up two or more operating units in deadly parallel, as discussed on page 133. A telephone company may create a series of nearly identical divisions, or finance companies may organize each of many offices on the same general pattern. Control is enhanced because the results from any one office may be compared against the performance in the others. This deadly-parallel arrangement removes a great deal of personal opinion in setting standards. As we mentioned in Chapter 3, it is important that employees accept standards as reasonable; if one branch meets a given standard, an aura of reasonableness is created for that standard and a wholesome attitude toward it tends to develop throughout all branches.

A proper grouping of activities can aid control in still another way. By placing together activities that are closely interdependent, we can reduce the amount of "overhead" control that is required. When interrelated work is done in several different departments, we have to control with precision the quality and flow of work as it moves from one department to another. Even with the best of controls a mistake is likely to result in arguments and buck-passing. So a more satisfactory arrangement is to assign the interrelated work to a single department or "project team," or to an individual.

Product divisions of a decentralized company are perhaps the best examples of this basic idea. Each product division is in charge of its own production, selling, engineering, and other essential functions. Key people in the various departments know one another and exchange information freely and informally. If production falls behind schedule, the sales manager probably knows it almost as soon as the production manager, and so the former adjusts sales efforts and delivery promises accordingly. Or if price competition is very keen on a particular item, the engineering and production men find out about it and adapt their activities with an eye to cutting costs. In short, the division functions as a team. Elaborate controls are not imposed from someone several organizational levels higher; instead, control information is promptly available to the people best able to do something about it. The pressures for corrective action arise from the situation, not from the arbitrary requirements of some high official.

In an organization where interdependent work is combined together in a single unit, supervisory control focuses on end results. Data on in-process activities do not pass through several layers of supervision but are fed promptly to the appropriate members of the team where they serve as a basis for self-regulation. Control is not only simplified, but there is also a much better chance for developing constructive attitudes toward control.

Profit Centers

"Profit centers" are a valuable control device if organization and controls are well matched. A product division with its own engineering, production, and marketing—to continue this example—can be judged in terms of the profit it earns. Since the profit figure shows the *net* result of all the activities within the division, its use as a control standard encourages coordinated and balanced effort. Also, executives within the division have to keep their activities in tune with the external environment in order to sustain profits. Although insensitive and slow to reflect intangibles, as already mentioned, the profit standard is the best comprehensive measurement that we have.

The temptation is to over-use the profit center idea. Some companies try to make each plant, each branch office, each warehouse, and even service units, such as purchasing, a so-called profit center. A profit is calculated for each unit. In effect, each unit buys its materials (often from other units), hires its own labor, and sells its products or services (perhaps to other units). Then after charges for overhead, a profit for the unit is computed. But how suitable is this resulting profit figure as a control standard?

Many profit center managers devote more energy to negotiating the artificial prices used to transfer goods in and out of their unit and the amount of overhead charged to them than they give to improving the activities which they can actually improve. Because of the profit control they spend a lot of unproductive time playing games with transfer prices. The trouble arises because the control measurement—profits—is much more comprehensive than the activities assigned to the unit they direct. Most plant managers, for instance, do not decide the specifications of what they make, how much to produce, nor whom to sell it to at selected prices. Consequently, control standards focusing on cost, quality, and delivery are more appropriate for such a plant manager than total profit.

Profit center control makes sense when (1) semi-autonomous, self-contained operating units are part of the organization structure, and (2) the primary objective of such a unit is profit. Managers of the semi-autonomous unit are free to adjust to new opportunities, and we want to encourage their initiative.[3] But we must be careful that the control directs that initiative to the result we want. If the main purpose of a district office, for example, is to build sales volume, we will confuse matters by calling the office a profit center. The control we select should reinforce the intent of the organization.

Adjusting Controls to Matrix Organization

Distinct control issues arise when a matrix form of organization is used. Increasingly popular because of its adaptability, a matrix organization has two types of units—resource departments which develop pools of specialized talent (and perhaps facilities), and project teams to which talent is assigned to work on specific projects. Law firms, management consultants, construction engineering, and

---

[3] Profit center control presumes independence of action. Local managers devise their own responses to their dynamic environment. This is in sharp contrast to the control approach discussed in Chapter 4, where we try to shield operating managers from external variables (by regulating operating conditions, standardizing products, etc.), and use "flexible budgets" and standard cost accounting to separate internal efficiency from external changes in sales and prices.

other professional service groups have used matrix organization for many years. Producers of space vehicles and missiles commonly adopt a matrix structure, and more recently other enterprises that want concentrated, coordinated attention on a particular venture will organize in this manner.

In a matrix organization, control is carried out primarily within each project team. The control tools normally will be those described in Chapter 5 for projects and programs. And the director of each team—the interim line boss—will be the person to balance the impact of the diverse controls so that the major mission of the team can be achieved. Selected reports on progress will be made to senior executives in charge of the respective team. In all these respects the control design is straightforward and quite in line with project and program controls already discussed.

The troublesome issue is what role the resource departments should play in controlling. For instance, in a shipbuilding company should an engineering department (or marketing, legal, purchasing, finance, or the appropriate resource department) actively participate in controlling the quality and amount of engineering work done on a new container cargo vessel? Or in a large legal firm, should the criminal law department control the briefs prepared by a young criminal lawyer assigned to a major anti-trust case? Having provided the needed personnel and perhaps other services, the resource departments take on the role of staff. The question then arises, while in this staff posture what control, if any, should the resource departments exercise?

Three control designs indicate the array of possible relationships.

1. *Full delegation to project teams.* In a decentralized matrix design each project team has virtually full independence in the action it takes. The presumption is that the uniqueness of the project, the need for prompt and coordinated action, and/or the high competence of the various specialists assigned to the team make localized control preferable. The team is on the field and it is up to the players to win the game.

Such independence relates to the running of specific projects. Post-action review by the resource department will, indeed, be carried out so that improvements in training for future projects can be made. These post-action reviews also seek to evaluate the effectiveness of persons assigned to the teams (insofar as this can be unscrambled from the group effort). Future assignments as well as advice on bonuses or other rewards will be influenced by this evaluation. Obviously, this post-action review is of personal concern to team members, so they are unlikely to disregard the training, suggestions, and values of their resource "home" department.[4]

2. *Periodic concurrence.* A second arrangement is for the resource department to participate in milestone reviews. As we saw in Chapter 5, these milestone reviews are typically yes-no control points when the progress-to-date is used as a basis for authorizing continuation of the project as planned.[5] Resource departments may join in such assessments and their concurrence may be necessary for the work to proceed.

All yes-no controls are introduced to improve safety, and adding an evaluation by resource departments increases the chances of catching serious errors. The chances for delay are also increased. So the desirability of requiring concurrence of resource departments at milestones depends on the relative weight attached to safety vs. speed and some expense. Much depends upon how provincial the resource department is in its standards; it is possible, though by no means common, for the resource department to be as customer-oriented and results-oriented as the project manager himself.[6]

Naturally, the people on the operating teams will try to anticipate possible objections by the resource departments (and whose views will be sustained if a difference of opinion is appealed to higher authority). This anticipation of control at the review points will temper actions in advance because prompt clearance simplifies the operating task. So, even though reviews are occasional, any re-

---

[4] The close analogy of matrix organization and military organization is apparent. In the Army the corps are the resource departments. They develop the pools of specialized talent; a field force with an operating mission is then made up of a combination of specialists supplied by the various corps. Likewise in control, the corps headquarters do not attempt to control action in the field, but they do evaluate the effectiveness of their protegées and do exercise a strong influence on future assignments of these people. Military history abounds with accounts of how individuals responded to the dual influences of their operating assignment (including the line commander) and their corps.

[5] Reasons for preferring milestones to annual reviews are discussed on page 69.

[6] The IBM corporate staff concern with operating division plans is akin to the control described above. In the IBM scheme, the staff department plays a lesser role as a resource center, and it is expected to take greater initiative in suggesting alternative courses of action. Also, provisions are made for holding the staff as well as the operating division accountable. IBM feels that the magnitude and importance of plans warrants the extra overview by the corporate staffs.

source department whose consensus is required exercises substantial influence.

3. *Continuous monitoring.*   A highly centralized control by resource departments is possible, usually by making the departmental representative on the project team merely a communicator of reactions and decisions made by the control staff group. This arrangement will undermine the project team concept. It can be tolerated only for those functions which make minor and intermittent inputs to the team's action—as may be true of, for example, a legal or real estate department. In effect, these fringe functions are withheld from the matrix. They perform an occasional service and exercise yes-no control at a few points when their specialty is involved.

Control in a matrix organization, then, does involve more or less double-checking by the project teams and their line supervisors and by the resource groups. The scope and frequency of the double-checking calls for a careful trade-off between greater freedom for the project team and more safety by closer "staff" scrutiny.

"*Product Managers*" pose similar questions of dual controls. Although the term "product manager" applies to a wide range of organizational designs, a central theme in all of them is that of a person who follows the fate of a product (or group of closely related products) throughout a company. He watches marketing, advertising, inventories, production, purchasing, research and development, and external supply and competition of his particular product with the aim of maintaining its contribution to company objectives. Unlike a matrix organization, however, the work of the various functions remains within the respective departments, and the product manager tries to cajole, seduce, persuade, or otherwise induce the respective departments to treat his product with wisdom and tender care.

Product managers have less power than do project team leaders in a matrix organization. In turn, project team leaders have less power than managers of self-contained product divisions. These three alternative ways of providing for product coordination and control are shown in Figure 10-2.

Under the product manager set-up the functional departments do most of the controlling since they do the operating. The issue is the extent to which the product manager also controls. Quite naturally, the functional departments tend to be provincial in their attention and values; also they normally have many products and

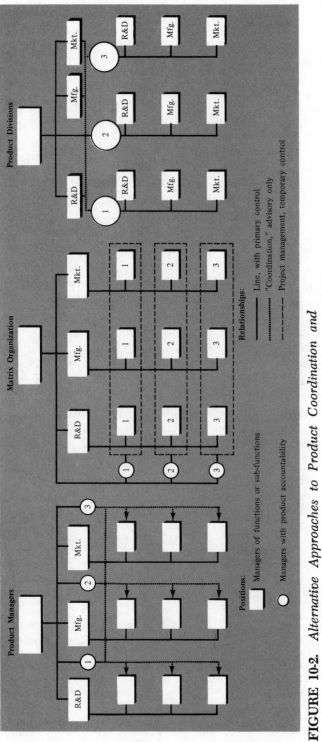

FIGURE 10-2. *Alternative Approaches to Product Coordination and Control.*

157

programs to serve. Consequently, if the program manager is to be effective, he should exercise some control over his product.[7]

The product manager should have steering-controls, yes-no controls, and post-action controls at his disposal. He cannot do his job well without them. But his is a staff role and restraints are in order.

1. With rare exceptions, he should obtain his internal control data from the same sources as the operating departments and should use mutually agreed-upon standards in evaluating results. As with any good control, the primary aim is to induce the operating people to exercise self-control. They should be watching the same indices as the product manager, though admittedly with less concentration and with less knowledge of related functions.

2. His corrective action should be predominantly in the form of mutually agreed-upon action; differences of opinion, if any, regarding key moves should normally be resolved by a common supervisor.

3. He should have concurring authority at a few yes-no control points dealing with such matters as price changes, product additions or deletions, and allocation of the sales promotion budget for his product. Again, differences of opinion can be appealed up the line.

With product managers, as with matrix organization, we retain both a functional and an end-product viewpoint in the organization and in the controls. Primary operations remain in the functional departments in one set-up and are transferred to project teams in the other. Accordingly, the primary control activity should be placed where operations are located. But in each configuration control data are also made available to persons cast in the staff role. The duality in organization is reflected in a duality of controls. However, the frequency of review and the way of exercising influence differ significantly, as already stated in the preceding paragraphs.

Matching Leadership Styles and Rewards with Controls

Plans, organization, and controls must be fitted together in a reinforcing structure. But to achieve full effect, leadership also should suit the management design.

---

[7] A comparable set of issues relates to a "program coordinator" or a "project coordinator." Sometimes, regional offices are involved and the relevant factors vary, but the concept of a staff coordinator working across a series of operating departments is the common feature.

How much permissiveness?   Leadership styles vary. In the wide range of styles from authoritative to counseling, McGregor, Likert, and others point to permissiveness as a key element.[8] Permissiveness here refers to the degree of freedom allowed subordinates to interpret instructions and to devise their own ways of meeting specific situations. Obviously, permissiveness affects control; as more local autonomy is exercised, predictability of actions declines. And in situations in which dependability and close coordination are vital, such a decline in predictability may be serious. On the other hand, if the technology permits permissive leadership, we gain in local initiative and voluntary coordination.

Since leadership style is intertwined with the application of control standards, there should be a compatibility between leadership style and control design. For instance, to couple a control system predicated on close observance of standards with a permissive leadership style invites trouble.

Our previous analysis of control design clearly indicates that some situations call for tight control whereas loose control is preferable in others. The same situational factors bear upon the choice of leadership style. Some trade-offs of benefits may be involved in the final selection of controls and leadership style, but it is vital that the controls support the leadership style, and vice versa.

Leadership style is not established by public announcement; it is based on executive behavior. Regardless of what a boss or a manual may say, those controls that are enforced are to the persons being controlled an unembellished guide to what they must do well and what they can do indifferently. They soon learn, for example, whether a "no smoking" rule means what it says or is merely a suggestion of desirable behavior. It is the action of the supervisor in disregarding or insisting that the standard be maintained that gives meaning to the control. Enforced standards communicate.

The permissive leader obviously faces a tough decision on which control standards he will seek to enforce. If he does not follow up on, say, established quotas, antipollution standards, or routine matters of attendance, his subordinates will infer that he is indifferent to their actions in such areas. This kind of permissive-

---

[8] D. McGregor, *The Human Side of Enterprise.* New York: McGraw-Hill Book Company, 1960, presents in the author's words his well known Theory X and Theory Y alternative approaches to leadership. See also R. Likert, *New Patterns of Management.* McGraw-Hill Book Company, 1961. Permissiveness is only part of their concern. For instance, they stress participation—as we do in setting pars and evaluating results. The text above concentrates on permissiveness because it is the source of greatest confusion between "modern" leadership ideas and constructive control.

ness results in no control. On the other hand, if he consistently checks up on deviations of actual performance from standard but then confines his response to mild suggestions for improvement, the amount of control will be a function of the motivation of the subordinate. Of course, even a permissive leader may single out a few subjects on which no permissiveness is tolerated; the key here is to pick these subjects carefully.

Consistent rewards and controls. The company reward system, like leadership style, should reinforce the controls. Every production man knows that a bonus based on volume alone leads to neglect of quality. Similarly, if professors get promoted on the basis of publications, their teaching suffers. Perhaps the most common error in management practice is to reward people for short-run results while urging them to take a long-run viewpoint; such short-run pay-off is particularly insidious because long-run results are hard to measure and control. With a recognized reward (or penalty) associated with one kind of result, even the best designed controls on other results will receive secondary attention.

Individuals respond to many kinds of rewards, tangible and intangible, on the job and off the job. Tying these rewards to controls is not simple. Only part of the rewards can be granted or withheld by management.[9] Often persons who take actions which lead to rewards—for instance, make promotions or provide challenging assignments—differ from the persons exercising control. The timing and occasions for rewarding action is often separated from evaluation of results, perhaps by several years. Actions viewed partly as a reward may be subject to other considerations—e.g., promotions.

Consequently, a careful review of the reward system should be made along with any major redesign of the control system. Insofar as flexibility permits, the granting of rewards should be clearly and explicitly related to desired performance as reflected by the controls. Usually the line supervisor will carry the primary part in this mating of rewards and controls. The interconnection is so important, however, that an occasional independent check by someone with an objective viewpoint is desirable.

[9] Widespread practice in governmental agencies of granting salary increases solely on the basis of years of service, coupled with heavy reliance in promotions on civil service examinations and other criteria unrelated to performance, sharply curtails management's ability to adjust rewards. As a result, response to controls is largely dependent on voluntary cooperation based on personal views of professional behavior.

Conclusion. Like the nervous system in the human body, control is only one of the vital subsystems in effective management. Planning, organizing, and leading are also essential. And all these subsystems interact. If we change one, we may need to redesign the others also.

This interaction is a potential source of strength. By designing an organization which is suited to company plans, and reinforcing both with compatible leadership and controls, we can generate a high synergistic force.

Several ways to obtain such an integrated management design have been flagged in this chapter. Sharpening plans so as to provide clear control targets, using control data for new planning, decentralizing without loss of control, creating departments and profit centers which aid control, adjusting control to matrix organization, and enhancing control by compatible leadership and rewards all illustrate the possibilities of reinforcement.

The design of good controls is an intriguing task. Fitting them neatly into a balanced management structure is even more challenging—and rewarding.

FOR FURTHER READING

Jerome, W. T., *Executive Control*. New York: John Wiley & Sons, 1961. Part V describes du Pont, General Electric, and Koppers approaches to a central control system.

Schleh, E. C., "Grabbing Profits by the Roots: A Case Study in 'Results Management,'" *Management Review,* July 1972. Shows need for close tie between planning, organizing, and controlling.

Steiner, G. A., and W. G. Ryan, *Industrial Project Management*. New York: The Macmillan Company, 1968. Analysis of work of sixteen successful project managers in aerospace industry.

Vancil, R. C., "What Kind of Management Control Do You Need?" *Harvard Business Review,* March 1973. Cautions on indiscriminate use of profit centers.

chapter 11

PUTTING AN IMPROVED SYSTEM TO WORK

In a dynamic world the design of a control system is never finished. New opportunities for service call for fresh programs, personal aspirations and habits shift, critical shortages are overcome and new ones arise, technology simplifies communication, even managers change. To keep abreast we must adapt our control mechanisms. Such progressive adjustment involves (1) redesigning various parts of the measuring and feedback systems we rely upon, and (2) gaining acceptance and skill in utilizing the improved systems.

Developing Sound Control Designs

Designing controls suited to particular purposes has been the central topic throughout this book. The recommended approach has three dimensions.

1. *Well-conceived individual controls.* A whole array of controls will be found in every organization. Some focus on past action; others are yes-no screening devices; but the most constructive influences on future behavior are steering-controls. Deciding which of these control types fits the particular situation is a good beginning. We can then turn to the specific design.

Each control cycle includes a series of elements. Since weakness of any element may undermine effectiveness, we should think through each control element by element. For a steering-control, the elements are desired results, predictors, composite feedback, pars, information flow, valuation, and corrective action. Guides for refining these elements are discussed in Chapter 2.

In addition to these rational, mission-focused aspects of control design, we must also consider the likely human response to each control. To reduce resistance and to elicit a constructive reaction, several steps are proposed in Chapter 3: Relate the control to meaningful and accepted goals of the people whose behavior we seek to influence; set tough but attainable pars; limit the number of controls and minimize competition for attention; confine detail control to self-adjustment; and develop a discerning view of measurements used.

This kind of analysis helps to sharpen each control and to give it behavioral viability. However, we have not yet dealt with the related issue of selecting the specific results that our well-formed controls are to promote. So we must turn to a second dimension.

2. *Controls for specific purposes.*   Companies, and departments within companies, vary in kinds of control they need. Their strategy dictates an optimum technology. The technology calls for a particular way of managing, and that style of managing creates the need for controls which serve very specific purposes. Although no single book can deal with all the diverse controls which might be needed, we have explored a representative array.

Repetitive operations need stability and dependability. So we single out repetitive elements for control, establish a normal satisfactory level of performance, and then build acceptable behavior patterns based on customary measurements and feedbacks. Dynamic action takes off from this dependable base.

Control is also needed for projects and programs. Here, Gantt charts and PERT networks help keep work on schedule, and related cost estimates provide a basis for controlling the use of scarce resources. As programs extend over longer periods, milestone resources and monitoring of planning premises aid in steering-control.

Conservation and proper use of resources—things, people, capital, etc.—pose still different control problems. The inevitable overlap of resource controls and operating controls generates tensions. So we have to distinguish between controls that assure adequate supply and controls over the use of resources. Supply controls can be separately administered, but use controls have to be carefully coordinated.

Creative work is especially difficult to control because its output is so uncertain. Nevertheless, controls have been devised for both R&D and advertising. In the short-run, periodic and milestone reviews keep creative effort directed toward desired results. Assessment of people, however, is delayed until a "batting average" based on a series of opportunities has been compiled. Since evaluations are largely subjective, group judgments are typically used.

Control of company strategy involves both long-range results and high uncertainty. Consequently, updating forecasts and monitoring the environment become important steering-controls. Control of the process of strategy formulation is also useful. But to tie controls to individual performance, a special application of reaching back is necessary.

Clearly, controls can be devised for a wide range of purposes, as the examples just cited (and described more fully in Chapters 4 through 8) show. The control cycle with its basic elements is present in each case, but factors watched and the feedback arrangements differ greatly.

3. *Balanced control systems.* The numerous controls devised to meet specific needs compete for attention. Unless we are careful, convenient controls on minor matters will override controls on vital but intangible results. Also, central coordinating controls must be sorted out from distinctly local controls. A single grand design is impractical, but the controls bearing on a division or a branch should be integrated into a balanced subsystem.

Since controls become effective only through the response of individuals, a good place to start designing systems is with the set of controls bearing on each key person. Each set should be matched against the total results that an individual is expected to achieve, and then balanced in terms of desired emphasis.

Next, as suggested in Chapter 9, these localized sets should be drawn together insofar as (a) coordinated effort is necessary, (b) significant economies in measuring or evaluating are possible, or (c) adequate attention on a vital factor must be assured. Also, some consolidated reports are necessary to tie local action to central objectives.

And the emerging control systems should be integrated with other managerial arrangements for planning, organizing, and leading. Perhaps the controls will have to be tailored, or maybe the other activities can be modified to aid the controlling in ways outlined in Chapter 10. The aim here is to obtain synergistic efforts from planning reinforcing control, organization reinforcing planning, and so forth.

Emerging from this total process will be control designs reflecting the needs of each particular situation. As the needs change so, too, will the optimum control design. If we have done our designing job well, the new designs will provide a positive managerial force, concentrated on behavioral responses, with a strong future orientation, and balanced in attention to intangible and long-run results.

Modifying Customary Behavior

Just as a blueprint has to be transformed to become a house, so diligent effort is required to convert a control design into a living reality. More than formal instruction is involved. Both adjustment of the social structure and modification of personal values are necessary.

An enterprise is productive only when it has its own social structure. To work together effectively, people need to know what to expect of others, what their own role is, where they can get help, and who has power. A new control design often upsets these established relationships.

Introduction of profit center control, for instance, typically reduces the power of resource divisions and raises questions about measuring the output of R&D work. Similarly, when a PERT control is installed, the people who receive the new detailed progress reports can dip into operations formerly managed entirely by a local superintendent. Even the addition of a pollution control officer and a new set of yes-no screening points will alter information flow and the customary path for engineering projects.

We do not know what such announced changes really mean until we have had actual experience with the new relationships, have observed carefully the behavior of new people, and have tested the strength of the central staff. The new social structure—the way of working together—takes time to form.

Values of individuals may also have to change to fit the new design. For example, when a large insurance company set quotas a few years back for employment of blacks, the conventional attitude of supervisors about necessary education had to be adjusted before the new standards were felt to be feasible. Clearly, if professional self-control is substituted for external control of, say, customer service or working hours, personal values must support the new arrangement.

So difficult are some of these value shifts that a new set-up is ineffective until persons compatible with the new way of life are put in key posts.

Allowing Time to Absorb the Change

Conversion to a new control design takes time. Social structure and personal beliefs cannot be altered overnight. Managers can assist in the transposition, however, by dealing with three psychological factors: learning, anxiety and confidence (see Figure 11-1).

Learning new relationships and attitudes, like other learning, is aided by clear explanations, opportunity to try the new way, further questions and explanations, more trials and adjustment, and then practice. If a manager helps everyone involved recognize that this kind of process may be tedious at first but will avoid confusion later, the total transformation will be expedited. But mature and successful people will not always willingly accept the need for learning.

Any change that alters a man's primary source of satisfaction for his security and self-expression needs is sure to create anxiety. Just the uncertainty about how a new control will affect him personally is unsettling. Such anxiety often causes undesirable behavior—irritability, resistance, lack of enthusiasm. A manager should do all he can to relieve anxiety during a transition period. Stating facts, explaining future plans, stressing future benefits, having people meet new associates, scotching rumors, showing awareness of a man's personal problems—all help allay anxiety. With rare exceptions, bad news faced promptly is better than extended worry. If answers to specific questions cannot be given, assurances of when and how the information will become available is helpful.[1]

Both learning and relief of anxiety help to rebuild confidence. In addition, a manager can bolster confidence by reinforcing desired behavior. Public recognition and reward to persons who successfully utilize the new control design transfer attention from old ways to the new pattern, and continuing acknowledgement of success restores a sense of competence that had been jeopardized when familiar behavior had to be altered.

The personal and social adjustments just described take time. Because of this required investment in time and energy, we naturally hope that a new structure can be used for several years. Like research for a new medicine or tooling-up for a new airplane model, we want a period of stability when we can recoup our investment. Similarly, most people need a spell of stable productivity following a siege of readjustment. Although we anticipate recurring need for change in control design, the wise manager knows that there are personal and economic tolerance limits to the frequency of change.

---

[1] Participation in designing the controls and setting pars will speed up learning. On the other hand, participation may extend the period of anxiety, unless the participant sees clearly that he will fare well in any alternative being seriously considered.

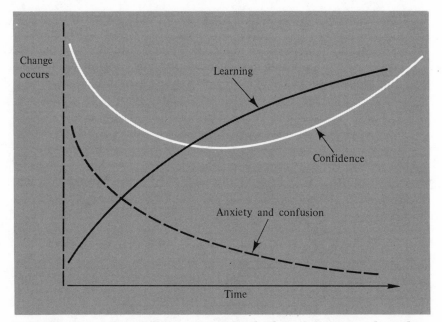

FIGURE 11-1. *Psychological Factors and Change. Diagram shows how psychological factors are affected by change.*

Good control design that is well absorbed into the social fabric of an organization has subtle benefits. Instead of being an irritating and repressive influence, control activity is regarded as a positive force. It helps us achieve accepted goals and to anticipate new opportunities. With effective, modern controls, we too can land on a moon.

FOR FURTHER READING

BARTLETT, A. C., and T. A. KAYSER, eds., *Changing Organizational Behavior*. Englewood Cliffs, N.J.: Prentice-Hall, Inc., 1973.
Useful articles by behavioral scientists. Most of the ideas can be applied to putting a revised control system into effect.

DEMING, R. H., *Characteristics of an Effective Management Control System in an Industrial Organization*. Boston: Harvard Graduate School of Business Administration, 1968. Case study of the way a comprehensive control system actually worked in one company.

NEWMAN, W. H., C. E. SUMMER, and E. K. WARREN, *The Process of*

Management, 3rd ed. Englewood Cliffs, N.J.: Prentice-Hall, Inc., 1972, Chapter 28.

Provides a framework for thinking through the interplay between company strategy and management planning, organizing, leading, and controlling.

INDEX